WHAT YOUR COLLEAGUES ARE SAYING...

This most recent addition to the *Figuring Out Fluency* series is outstanding! It provides guidance and support for building foundational fluency skills and practical ways to implement meaningful assessment and joyful practice. This book could shift mindsets about what it truly means to be fluent in math.

Deborah Peart
CEO & Queen Mather of My Mathematical Mind
Ocala, FL

SanGiovanni, Bay-Williams, and Katt have done it—the ultimate fluency resource! The focus is on helping students through their struggles using the lens of the foundational knowledge and skills needed for computational fluency. This is a perfect blend of lesson seeds, routines, games, and centers that provide first instruction or intervention for every student. This book, coupled with self-reflection, humility, and productive struggle, provides a rock-solid foundation so all students can weather any storm!

Ron Perry
K–4 Math Specialist, Heim Elementary, Williamsville Central School District
Williamsville, NY

Figuring Out Fluency—Ten Foundations for Reasoning Strategies With Whole Numbers is a game changer! In this companion to the *Figuring Out Fluency* series, the authors provide teachers with hands-on, engaging resources to support teaching and learning of foundational fluency concepts. This resource uses games, routines, and centers to help focus on reasoning about fluency concepts!

Latrenda Knighten
Mathematics Curriculum Supervisor, East Baton Rouge Parish School System
Baton Rouge, LA

Undoubtedly the most impactful book in the series, *Figuring Out Fluency—Ten Foundations for Reasoning Strategies With Whole Numbers* is a must-have for every elementary teacher. It explains not only the foundational skills and concepts needed for fluency but also the how and why they are necessary for strategic thinking. The look-fors and practical, easy-to-implement ideas for instruction and practice will help students build their mathematical identities and grow as mathematical thinkers.

Brenda Dzwil & Holley Duffy
Instructional Coaches, Newington Public Schools
Newington, CT

Get ready for a comprehensive yet accessible read for teachers to lay the roadmap for equipping students with indispensable foundational skills for learning fluency strategies. The authors outline ten foundations and supplement them with descriptions, examples, and teacher-friendly language to implement in the classroom tomorrow. The variety of included activities will create a book filled with dog-eared pages that teachers will reach for regularly.

Marci Ostmeyer
Professional Development Director, Educational Service Unit 7
Columbus, NE

The ten foundations for reasoning strategies shared in this book are practical and designed for immediate classroom application. They encompass methods to teach, practice, and assess fluency, which foster a supportive learning environment that cultivates a positive math identity among students.

Janel Marr
Math/STEM Resource Teacher, Windward District
Kānéohe, HI

This book provides teachers with an understanding of how to support and solidify foundational skills for students. It gives teachers a toolkit of assessments, games, and centers that are easily accessible and effortless to implement. It also challenges the traditional classroom, pushing the teacher to become more of the facilitator while encouraging an equitable education for all students.

Marissa Giangrosso
Instructional Coach
Blue Springs, MO

I own every book in the *Figuring Out Fluency* series, and this is the missing piece. This book examines the why and how number sense works in building procedural fluency with tools both for student practice and assessment of understanding. The structures and strategies are explained in a way that new and veteran teachers can implement in the classroom.

Christina Worley
K–5 Math Curriculum Developer, St Lucie Public Schools
Port Saint Lucie, FL

Figuring Out Fluency—Ten Foundations for Reasoning Strategies With Whole Numbers at a Glance

Building off of *Figuring Out Fluency*, this classroom companion explores the foundational ideas that are essential to numerical reasoning and number sense.

FIGURE 7 ● Foundations for Computational Fluency

FOUNDATIONS MODULES	WHAT IT IS (IN BRIEF)	EXAMPLE PROMPT
1. Number Relationships: Comparison and Estimation	Knowing approximately where a number is in relation to other numbers	Please place 28 on a number line. Use various endpoints [0, 50], [20, 30], or [0, 100].
2. Subitizing and Decomposing	Subitizing is visually seeing subsets of the whole to determine the whole; decomposing is starting with the whole and determining the parts	How many dots? How do you see them? **Decomposing:** There are 10 turtles on two logs. How many might be on each log?
3. Distance to 10, 100, and 1,000	Seeing how far a number is from a benchmark (over or under)	How far is ... ● 8 from 10? ● 37 from 40? ● 107 from 100? ● 288 from 300?

Overviews of each foundation show how it contributes to fluency strategies, how to assess it, and how to teach it explicitly either as first instruction or intervention.

MODULE 1

Number Relationships: Comparison and Estimation

NUMBER RELATIONSHIPS OVERVIEW

In mathematics, students must understand and compare quantities. Comparing quantities involves understanding the relative size of a number; for example, knowing that 8 is 1 more than 7, which also means 7 + 1 = 8 and 8 – 1 = 7. This grows into noticing that 47 is 3 away from 50 and 6.92 is 8 hundredths away from 7. Importantly, comparing is not about using the < and > symbols correctly; rather it is knowing that 8 is more than 7 because it is one more object and it is the number to the right on the number line. Estimation is similar to comparison but allows for approximation: for example, noticing that 47 is close to (but to the left of) 50. Students need to be able to find the approximate location of a number on a number line as it relates to benchmarks. Understanding these two number relationships develops students' number sense. Together, they lay the foundation for estimating sums and differences, which is the focus of Module 10.

In this module, the focus is on placing numbers on number lines, changing their locations as endpoints change, and reasoning about how close numbers are to one another. For many students, this work is often procedural. Comparing numbers is consigned to looking at place value. Students may have limited work with number lines, and the experiences they do have may imply that number lines are static—always having the same endpoints (e.g., 0 and 100). Connecting estimation to comparison and number line work is challenging because estimation doesn't typically play a large role in mathematics classrooms. Also, students can misinterpret what it means to estimate, thinking that there is a correct estimation. Perceiving one right way to estimate leads to inflexible, procedural approaches to their reasoning.

HOW DO NUMBER RELATIONSHIPS CONTRIBUTE TO FLUENCY STRATEGIES?

Understanding numbers and their relationships is inseparable from computational fluency. It helps students select a strategy, execute it, and determine if their results are reasonable. In 48 + 25, a student recognizes that 48 is close to 50 and they can use the Make Tens strategy. They break 25 into 2 and 23, giving 2 to 48 to make 50. From there, they add 23, finding the sum of 73.

$$48 + 25$$
$$2 \overset{\frown}{+} 23$$
$$48 + 2 + 23$$
$$50 + 23 = 73$$

This also aids in determining reasonableness. This same student can reason that 48 is close to 50 and that 25 is close to 30. They determine that 50 + 30 = 80 so then the sum of 48 + 25 should also be close to 80.

A strong understanding of number relationships also helps students represent their thinking, which is especially useful as numbers become larger or more complex. In both examples, the students manipulate their number lines for convenience. Neither number line uses endpoints of 0 and 1,000.

Students choose good starting places; they know how the numbers change as you move to the right or left. They can count in chunks (Module 5).

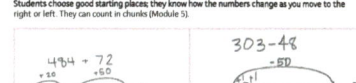

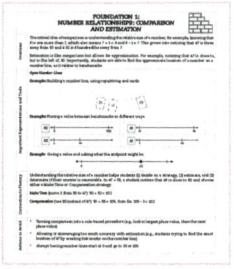

The number relationships reference page provides an overview, important representations, connections to fluency, and actions to avoid. It can be downloaded and used for reference in planning, teaching, and discussions with colleagues and families.

Available for download at https://online.qrs.ly/pcf6a5o

ASSESSING NUMBER RELATIONSHIPS

The goal of assessing student thinking related to number relationships is to see if students have a relational understanding of numbers. Here, we share observation ideas (*look-fors*) and interview or journal prompts (*quick assessments*).

Each foundation module includes teaching activities that help you explicitly teach the strategy.

ACTIVITY 1.2
NEAR AND FAR

Estimating helps students think about the *relative position* of numbers—how close or how far numbers are from one another. This allows students to think about numbers that are close to one another and those that aren't. By doing so, students will develop a better sense of what might be a good estimate: 50 is a better estimate for 45 than 80.

For this activity, gather a string and index cards. The string can be stretched out between two chairs by securing the ends or it can be attached to a wall or board. This string will serve as a number line where students will place number cards. Prepare number cards by folding each card in half and writing a number on each side (you may have heard this referred to as "clothesline math"). The numbers can be within any range that is appropriate for the students.

Place one card's fold over the string, so it hangs and shows the number. Start with a decade number—one ending in zero. Hold up another card and have students determine if it would be near or far from the first number. They will place it on the number line accordingly and provide their thinking about why they placed the card where they did. It might be easier for students to begin with numbers that are relatively close to each other. This allows students to visualize a number line. Shown in the following example, a 10 is placed on the string. Then students determine that 19 is far from 10, so this card is placed on the other end of the string. The next card is shown, a 12, and students determine that 12 is near 10. It is probably best to only place three or four cards at a time on the line. Otherwise, it might become too crowded with cards, which would make it difficult for students to think about the position relative to the initial number.

Continue to place different numbers on the number line and ask students, "Is this number near or far from the number that is on the number line?" Vary the position of where you hang the first number cards on the line, as this will encourage students to visualize a number line and think about the distance between numbers. It is important to note, however, that the exact spacing between the cards doesn't really matter for this activity. Some students may worry about the spacing and try to get it accurate, but that isn't the goal of this activity. Rather, it is to help them think about how close together or how far apart numbers are.

As a follow-up to a whole-class activity or station activity later, students can replicate this activity with partners. Students use a whiteboard to draw a line similar to the physical one used previously. One student places a number along the line and then gives the other student a number to position. The student who places the number says, "___ is near/far from ___." So, if a student writes 40 on the number line and offers the number 87, the second student places 87 further to the right of 40 and says, "87 is far from 40."

ACTIVITY 4.6

Name: "The Count" **Type:** Routine

About the Routine: This quick routine can give you insight into students' proficiency with counting while providing a good opportunity for practice. The Count is a routine that has students estimate and skip count. It is a good opportunity for practicing skip counting by a variety of intervals, which is essential for using the Count On and Count Back strategies. You can extend the routine by ending with a problem connected to the counts. For example, you can have students start with 43 and count on by ones. Then, have them start with 43 and count on by tens. Pose a problem, such as 43 + 48. Have students turn and talk with their classmates about how they can count on by tens and ones to find the sum.

> **TEACHING TAKEAWAY**
> You can increase engagement with routines by turning them into challenges or games. For The Count, you can have students write down the number

Mater
(optio

Direct

ACTIVITY 4.10

Name: 300 Is Perfect **Type:** Game

About the Game: This *300 Is Perfect* game blends estimating and counting. The goal of the game is simple: Players try to get as close to 300 as possible without going over. Each player starts at 0 and rolls a digit. That digit can represent a number of ones or tens. The player decides which it will be based on the number they are counting on from and the goal of getting close to 300 at the end of their eight turns. Along the way, they practice counting on by tens or ones from a variety of numbers. A good way to introduce the game is to set a goal of 100. But such a low target limits counting by tens. You can adjust the game to any start number and any target, such as starting at 200 and playing *500 Is Perfect*, or starting at 400 and playing *3,000 Is Perfect* (in this version, the digit can represent tens or hundreds). Some variations are shown in the game board examples.

Materials: A *300 Is Perfect* game board (optional), a 10-sided die or digit cards

Directions: 1. Players take turns rolling a digit.

2. The player decides whether the digit will represent the number of ones or tens.

3. The player begins at 0 and counts on the amount that they determined.

4. On the next turn, the player rolls again and decides how to use the number.

5. They count on that amount from the new number.

6. The player must roll exact

7. The player closest to 300

The image on the left shows an example you could modify it for a different starting by hundreds and tens with a goal of 3,000.

300 Is Perfect

Directions: Start at 0. Roll a number. Decide if that is the number of ones or tens to count on from it. Roll again and count on from your tens number. Get as close to 300 as possible without going over.

Tens	Ones	Total
6		60
	9	69
7		139
6		199
	4	203
	7	210
5		260
	8	268

This resource can be download

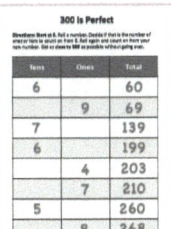

ACTIVITY 10.14

Name: Just One, Which One **Type:** Center

About the Center: This center is an independent practice to reinforce the notion of estimating a solution by adjusting one number in a problem (see Activity 10.4). It's a concept that students will need to practice, especially if their estimation experiences have been mostly procedural or focused on finding two benchmark numbers.

Materials: 10-sided dice, playing cards (queens = 0, aces = 1; remove tens, kings, and jacks), or digit cards; recording sheet (optional)

Directions: 1. Students generate a problem with dice or cards. Alternatively, you can provide problem cards written on index cards or printed on paper.

2. Students record the problem.

3. Students estimate the result by changing one number to a benchmark number and record the estimate.

4. Students estimate the result a second time by changing the second number.

5. Students identify which problem was easier for them to think about.

The recording sheet is shown. Students could fold a piece of paper in thirds, draw columns on a dry erase board, or write in their mathematics journals. Notice the extension added at the bottom of the recording sheet, asking them to explain whether it's easier to estimate by adjusting one or both numbers. You can swap that out with other prompts such as, "Tell me which problems were easier to estimate and why" or "Were there any problems that were easy for you to estimate if you changed either number? Which problems were they? What made them easier?"

Just One, Which One

Directions: Generate a problem and record it in the first column. Estimate the result by changing one number. Then estimate again by changing the other number. Star the one that's easier to think about.

Problem	Estimate #1	Estimate #2
77 + 57	80 + 57 = 137 ★	77 + 60 = 137

What are you noticing about which number to change to estimate?

This resource can be downloaded at **https://qrs.ly/psf6a5o**

Routines, Games, and Centers for each foundation offer extensive opportunity for student practice.

Download the resources you need for each activity at this book's companion website.

Downloadable Foundation Briefs are quick-shot reminders that can be referenced while planning, teaching, and in discussion with colleagues and families.

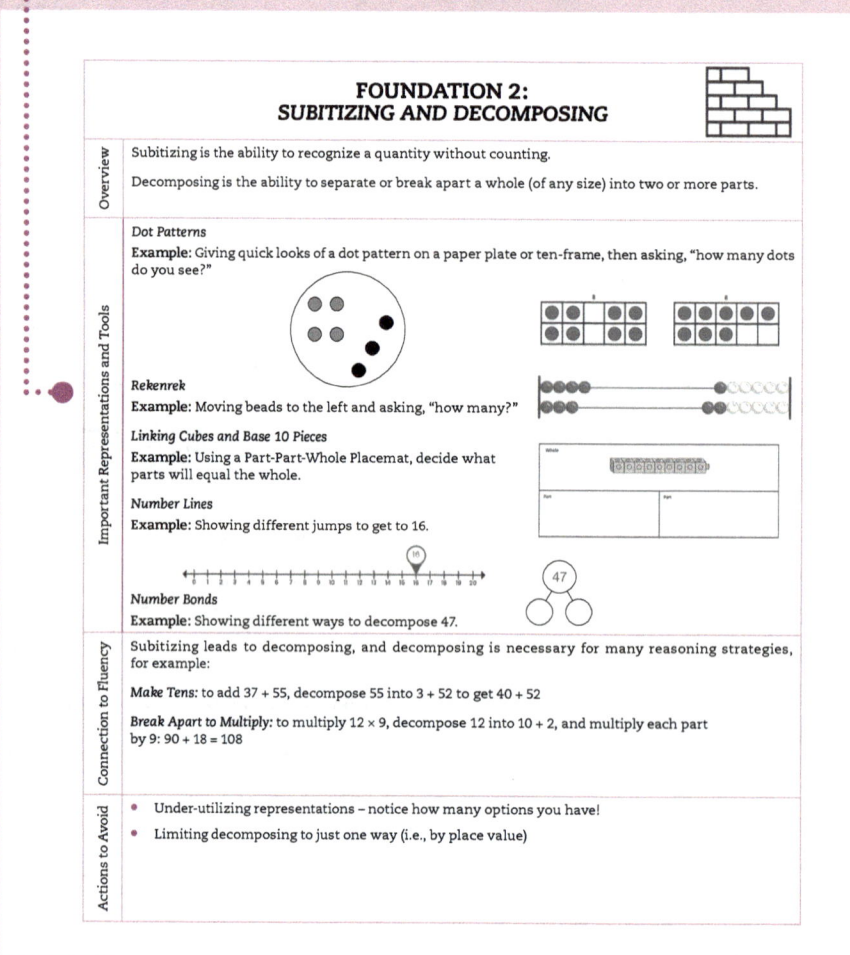

FOUNDATION 2:
SUBITIZING AND DECOMPOSING

Overview

Subitizing is the ability to recognize a quantity without counting.

Decomposing is the ability to separate or break apart a whole (of any size) into two or more parts.

Important Representations and Tools

Dot Patterns
Example: Giving quick looks of a dot pattern on a paper plate or ten-frame, then asking, "how many dots do you see?"

Rekenrek
Example: Moving beads to the left and asking, "how many?"

Linking Cubes and Base 10 Pieces
Example: Using a Part-Part-Whole Placemat, decide what parts will equal the whole.

Number Lines
Example: Showing different jumps to get to 16.

Number Bonds
Example: Showing different ways to decompose 47.

Connection to Fluency

Subitizing leads to decomposing, and decomposing is necessary for many reasoning strategies, for example:

Make Tens: to add 37 + 55, decompose 55 into 3 + 52 to get 40 + 52

Break Apart to Multiply: to multiply 12 × 9, decompose 12 into 10 + 2, and multiply each part by 9: 90 + 18 = 108

Actions to Avoid

- Under-utilizing representations – notice how many options you have!
- Limiting decomposing to just one way (i.e., by place value)

FIGURING OUT

Fluency

TEN FOUNDATIONS

for Reasoning Strategies
With Whole Numbers

A Classroom Companion

FIGURING OUT
Fluency

TEN FOUNDATIONS
for Reasoning Strategies With Whole Numbers

A Classroom Companion

John J. SanGiovanni
Jennifer M. Bay-Williams
Susie Katt

CORWIN Mathematics

For information:

Corwin
A SAGE Company
2455 Teller Road
Thousand Oaks, California 91320
(800) 233-9936
www.corwin.com

SAGE Publications Ltd.
1 Oliver's Yard
55 City Road
London, EC1Y 1SP
United Kingdom

SAGE Publications India Pvt. Ltd.
Unit No 323-333, Third Floor, F-Block
International Trade Tower Nehru Place
New Delhi – 110 019
India

SAGE Publications
Asia-Pacific Pte. Ltd.
18 Cross Street #10-10/11/12
China Square Central
Singapore 048423

Vice President and Editorial Director:
 Monica Eckman
Associate Director and Publisher,
 STEM: Erin Null
Senior Editorial Assistant:
 Nyle De Leon
Production Editor: Tori Mirsadjadi
Copy Editor: Erin Livingston
Typesetter: Integra
Proofreader: Jen Grubba
Indexer: Integra
Cover Designer: Rose Storey
Marketing Manager:
 Margaret O'Connor

This book is printed on acid-free paper.

24 25 26 27 28 10 9 8 7 6 5 4 3 2

Contents

Visit the companion website at
https://qrs.ly/psf6a5o
for downloadable resources.

Preface

All students can develop procedural fluency, including fluency with basic facts, whole number operations, rational number procedures and operations, proportions, and solving algebraic equations.

A COMMITMENT TO FLUENCY

To ensure every student develops fluency, we must first

- understand what procedural fluency is (and what it isn't),
- respect fluency, and
- plan to explicitly teach and assess foundations for computational fluency.

If you have read our anchor book, *Figuring Out Fluency in Mathematics Teaching and Learning*—which we recommend in order to get the most out of this classroom companion—you'll remember an in-depth discussion of these topics. In fact, Chapter 3 of that book was titled "Good (and Necessary) Beginnings for Fluency." This book, therefore, is a more in-depth exploration of that topic—delving into those foundational skills that underlie all fluency work, whether it is for first instruction, practice, or intervention. Since publishing *Figuring Out Fluency in Mathematics Teaching and Learning* in 2021, we have realized that this topic needed and deserved a fuller discussion and more tools for you as a teacher than could fit in a single chapter. We came to understand that all the good reasoning strategies in the world don't really help students if there are not some key ideas and skills solidly in place for them. Read on and we'll explain what we mean.

WHAT PROCEDURAL FLUENCY IS AND IS NOT

Before we get into the heart of this book, it's important to briefly discuss what fluency *is* and what it *is not*. Similar to fluency with language, wherein you decide how you want to communicate an idea, fluency in mathematics is a decision-making process: As you look at the numbers in the problem, certain strategies make the most sense. Thus, fluency is about having flexibility and efficiency. If a student can "fluently add two-digit numbers," they will likely add 49 + 48 and 51 + 14 differently; for example, they might add 49 + 48 by thinking they are both close to 50, so they would mentally round up by adding 1 to 49, adding 2 to 48 (so 3 total), then add 50 + 50 and subtract out 3. It equals 97. The other problem might be solved by adding tens and ones or using a Jump-Up strategy. Each of these options are efficient and are a good fit for the problem. In neither case is a standard algorithm a good option. It is slower and involves more steps. Thus, being fluent is not equivalent to being adept at using an algorithm. That is called, well, being skillful with an algorithm. Fluency includes being adept at algorithms, but that is only one of a repertoire of reasoning strategies one can use, depending on the situation. Having this fluency requires strong foundations in number relations, decomposing, and the properties. Fluency is not about speed, though it is about efficiency. Take for example, the problem 403 – 299. A fluent student will take the time to first analyze the problem, think of a reasonable method, notice that both numbers are close to hundreds, and use that information to solve it efficiently (for example, a Count Up strategy or a Compensation strategy). A non-fluent student with skill using the standard algorithm will stack, regroup, and solve, moving quickly through the steps. This student is fast but not efficient, as the algorithm will take longer and involves more steps. The National Council of Teachers of Mathematics (NCTM) Procedural Fluency Position Statement (2023) is an excellent description not only of what procedural fluency is but also what is necessary to ensure all students develop procedural fluency.

RESPECT FLUENCY

We are strong advocates for conceptual understanding. We all must be. But there is not a choice here. Procedural fluency relies on conceptual understanding, but conceptual understanding alone cannot help students fluently navigate computational situations. They go together and must be connected. The foundations in this book connect conceptual understanding to the essential skills needed for employing reasoning strategies with the operations and beyond.

EXPLICITLY TEACH AND ASSESS FOUNDATIONS

We cannot overstate the importance of readiness in developing fluency. Readiness for procedural fluency means developing the concepts and skills needed in reasoning. In numerous research reports, using learning progressions or trajectories are found to have a positive impact on student learning. As an example, consider the readiness to apply the Making 10 reasoning strategy. Students are ready to learn this strategy when they understand these foundations:

- Commutative Property

  For 5 + 8, the thinking begins with an add on to the larger number (8 + 5).

- Distance to 10/Combinations of 10

  8 is 2 away from 10.

- Decomposing

  Decompose 5 so that it is 2 and some more (2 + 3).

- Associative Properties

  Reassociate the 2 from the 5 with the 8 [8 + (2 + 3) = (8 + 2) + 3].

- Adding 10 and Some More

  Add 10 + 3.

This may look like a long list, but these are number relations and concepts that become automatic with adequate time and experiences. It is simply broken out here to illuminate how important foundations are to being able to apply reasoning strategies.

So, what are the significant reasoning strategies for which we need these foundations? In *Figuring Out Fluency in Mathematics*, we introduced seven such strategies for computational fluency:

1. Count On/Count Back (Addition and Subtraction)

2. Make Tens (Addition)

3. Use Partials (Addition, Subtraction, Multiplication, and Division)

4. Break Apart to Multiply (Multiplication)

5. Halve and Double (Multiplication)

6. Compensation (Addition, Subtraction, and Multiplication)

7. Use an Inverse Relationship (Subtraction and Division)

Notice that while there are seven strategies in the complete list, there are no more than four for any particular operation. So, for a child to be fluent in subtraction, for example, they would understand, be able to use, and know when to choose each of the strategies that are useful for subtraction: Count Back, Use Partials, Compensation, and Use an Inverse Relationship (i.e., Think Addition/Count Up).

To be ready to learn these different reasoning strategies, though, we must teach and assess certain critical foundations. The alternative (teaching Make Tens when students don't have these necessary foundations) results in students not being able to enact the strategy, and thus they are stuck using a counting method or memorizing their facts. They haven't learned a significant strategy that is incredibly useful for computation with whole numbers and rational numbers. In other words, we have not provided students the necessary opportunities and experiences to develop fluency. In this book, then, we focus on those ten foundations that are necessary for developing fluency. We have created a module for each one so that teachers have a plethora of teaching, practicing, and assessing ideas to ensure students understand and are adept at using each foundation. They include the following:

- Number Relationships: Comparison and Estimation (Module 1)

- Subitizing and Decomposing (Module 2)

- Distance to 10, 100, and 1,000 (Module 3)

- Counting and Skip Counting (Module 4)

- Properties of Addition and the Inverse Relationship with Subtraction (Module 5)

- Properties of Multiplication and the Inverse Relationship with Division (Module 6)

- Multiplying by Tens and Hundreds (Module 7)

- Multiples and Factors (Module 8)

- Doubling and Halving (Module 9)

- Computational Estimation (Module 10)

USING THIS BOOK

This book can support your curriculum and other resources, adding to your collection of high quality, student-centered activities. Fluency foundations take time and repeated experiences to develop, so this book can be thought about as a pantry—open it up when you need an assessment prompt to gain insights into your students' thinking (assess); you are hoping for some ideas on stories and visuals that build the foundations (teach); or you need an engaging routine, game, or center to focus on a selected foundation (practice).

As mentioned, this book is a classroom companion book to *Figuring Out Fluency in Mathematics Teaching and Learning*. In that anchor book, we lay out what fluency is and barriers to a true focus on fluency, and we briefly discuss necessary foundations for fluency. We also propose the following:

- 12 "fluency fallacies" that clarify what fluency is and how to accomplish it

- 7 significant strategies across the operations (previously listed), all of which require these ten foundations

- 8 "automaticies" *beyond* automaticity with basic facts, several of which are addressed in this companion book (e.g., decomposing [breaking apart] numbers within ten, doubling, and halving)

- 5 ways to engage students in meaningful practice, including routines, games, and centers

- 4 assessment options that can replace (or at least complement) tests and that focus on real fluency

- 6+ ways to engage families in supporting their child's fluency

In Part 1 of this book, we highlight some of these big ideas to provide context for the fluency foundation modules. Part 1 is not a substitute for the anchor book but rather a brief revisiting of central ideas that serve as reminders of what was fully illustrated, explained, and justified in *Figuring Out Fluency in Mathematics Teaching and Learning*. Hopefully, you have had the chance to read and engage with that content with colleagues first, and then Part 1 will help you think about those ideas as they apply to the ten foundational ideas in this book.

Part 2 is focused on modules about teaching, practicing, and assessing each foundational idea. Each module includes the following:

- **Overview:** an overview for your reference and to share with students and colleagues

- **Assessment:** a list of prompts related to the foundation, along with a variety of quick checks to determine what a student knows related to the foundation

- **Explicit Instruction**: a series of teaching activities that incorporate manipulatives, representations, and student talk to help students make sense of the foundation

- **Quality Practice**: a series of practice activities, including routines, games, and center activities that engage students in meaningful and ongoing practice to develop proficiency with the foundation while also serving as further opportunities to assess student learning

Note that most of the online resources include variations of the activities using numbers through the thousands place and decimals.

Pick and choose from Part 2! If these foundations are in your curriculum, then simply find the module that fits your needs and select the activities that best meet the needs of your students. Two of the needs you may identify are first instruction and intervention.

First Instruction: If your students are learning a fluency-related topic—for example, subtracting whole numbers or finding equivalent fractions—then ask yourself what foundations will be important to their success. Go to that module and select an assessment idea or practice activity to try out as a way to gain insights into students' readiness. You can immerse your students in a module, spending weeks exploring the foundation in depth, or you can identify a few activities to implement from time to time to ensure students maintain these necessary foundational skills. None of this has to happen all at once; activities can be woven into your instruction regularly over time.

Intervention: If you are providing intervention, then you are still picking and choosing content from across the modules. Looking through the book's table of contents, you might notice *decomposing* and wonder if the children you are working with are able to decompose numbers. Thus, you find an activity from this section and engage them in the task while you observe the extent to which they are able to decompose: Do they require manipulatives? Do they see all the ways or only some of the ways? Are they automatic? If the answer to the last question is "No," then continue to provide experiences for the child(ren), choosing more concrete activities if needed and moving into more abstract activities. If the answer to the automatic question is "Yes," then find another activity from another module and repeat.

Part 3 is about implementation! A series of FAQs are provided for implementation in various settings—the classroom, intervention setting, and at home. In addition, more ideas are provided for monitoring student success and ensuring that we continue to attend to students' emerging mathematics identities and agency as we engage in developing strong foundations.

WHO IS THIS BOOK FOR?

With over 120 instructional activities, 60 assessment prompts, and a companion website with resources ready to download, this book is designed to support many audiences, including classroom teachers, special education teachers, mathematics interventionists, tutors, and parents. The brief explanation of the foundation, suggestions for explicit strategy instruction, and range of options for practice make this resource useful for whole-class instruction, one-on-one instruction, and enjoyable mathematics at home. Additionally, those who lead teacher preparation programs can use this book to galvanize preservice teachers' understanding of the foundations necessary for fluency and provide these emerging teachers with a wealth of classroom-ready resources to use during internships and as they begin their career.

Figuring Out Fluency—Ten Foundations for Reasoning Strategies With Whole Numbers, A Classroom Companion is one of six companion books in the complete *Figuring Out Fluency* series. Each of these companions offers over 100 activities to support student reasoning related to different operations and types of numbers. We think of this book as sort of "Book 1.5." It is a useful in-between text between *Figuring Out Fluency in Mathematics Teaching and Learning* (the anchor book) and the other classroom companions:

Figuring Out Fluency—Addition and Subtraction With Whole Numbers

Figuring Out Fluency—Multiplication and Division With Whole Numbers

Figuring Out Fluency—Addition and Subtraction With Fractions and Decimals

Figuring Out Fluency—Multiplication and Division With Fractions and Decimals

Figuring Out Fluency—Operations With Rational Numbers and Algebraic Equations

Acknowledgments

As an author team, we know that each project is a collaborative effort with so many to thank. We thank the Corwin team (yet again) for making this book a reality. There is no better publishing team to work with than the Corwin Mathematics team. We are lucky to be a part of that team. We offer a special thanks to Nyle DeLeon for all her work on this project and to editor and publisher Erin Null for helping us cross the finish line. It is Erin's enthusiasm, partnership, insight, and friendship that makes this work so fulfilling.

From John: Thank you to the many mathematics educators and friends for your support, perspective, and friendship. I am lucky to have so many. You make me better. Thank you (again), Skip Fennell and Kay Sammons, for opportunity and mentorship through the years. Today, I am lucky to call you friends. Thank you, Jenny and Susie, for your passion, collaboration, insight, and effort. Thank you for appreciating my humor, among other things. I thank my wife, Kristen. There are simply no words to express my gratitude and appreciation.

From Jennifer: I am grateful to so many for their constant support, but I would like to acknowledge my mother, Joan, who passed away during the time we were working on this book. My entire life she supported and pushed me in every way, but mostly to be a strong and caring human generously serving the community. To the many students and teachers in Louisville and beyond who have shared their mathematics reasoning with me—thank you! To John and Susie, I am deeply grateful to you for collaborating on this book and working together to figure out the fluency foundation activities and experiences that set students up for success with fluency!

From Susie: I would like to wholeheartedly thank my family, Jason, Tenley, and Huxton, for being supportive of my "extra" work. I thank my dad, who reminds me of the value of commitment and new experiences. I express sincere gratitude to my close friends who cheer me on and are always there when needed. I would like to say thank you to Jenny and the Corwin team for allowing me to be part of this important work on fluency. Finally, I would like to thank John for inviting me to collaborate on another project and for sharing a strong passion for the work of supporting teachers and students in elementary classrooms.

PUBLISHER'S ACKNOWLEDGMENTS

Corwin gratefully acknowledges the contributions of the following reviewers:

Mary Duden
Lower School Math Specialist, Oregon Episcopal School
Portland, OR

Delise Andrews
Math Coordinator, Grades 3–5 Lincoln Public Schools
Lincoln, NE

Megan Jefferson
Math Specialist, Howard County Public Schools
Hanover, MD

Crystal Lancour
Supervisor of Curriculum & Instruction, Colonial District
New Castle, DE

Nicole Rigelman
Professor, Portland State University Education Program Officer
Portland, OR

About the Authors

John J. SanGiovanni is a mathematics coordinator in Howard County, Maryland. There, he leads mathematics curriculum development, digital learning, assessment, and professional development. John is an adjunct professor and coordinator of the Elementary Mathematics Instructional Leadership graduate program at McDaniel College. In addition to this *Figuring Out Fluency* series, some of his many Corwin books include *Daily Routines to Jump-Start Problem Solving, Grades K–8*; *Answers to Your Biggest Questions About Teaching Elementary Math*; the *Daily Routines to Jump-Start Math* series; and *Productive Math Struggle: A 6-Point Action Plan for Fostering Perseverance*. John is a national mathematics curriculum and professional learning consultant who also speaks frequently at national conferences and institutes. He is active in state and national professional organizations, recently serving on the board of directors for the National Council of Teachers of Mathematics (NCTM) and on the board of directors for NCSM.

Jennifer M. Bay-Williams is a professor of mathematics education at the University of Louisville, Kentucky, where she teaches preservice teachers, emerging elementary mathematics specialists, and doctoral students in mathematics education. She has authored over 40 books and 100 journal articles/book chapters, many of which focus on procedural fluency and developing mathematical proficiency. Beyond the *Figuring Out Fluency* series, these include *Math Fact Fluency*, *Everything you Need for Mathematics Coaching*, and *Elementary and Middle School Mathematics: Teaching Developmentally*. Jennifer's national leadership includes the National Council of Teachers of Mathematics (NCTM) board of directors and the TODOS: Mathematics for All Board of Directors and as president and secretary of the Association of Mathematics Teacher Educators (AMTE).

Susie Katt is the K–2 Mathematics Coordinator in Lincoln, Nebraska, where she leads professional learning, assessment, and mathematics curriculum development. She is a coauthor of *Productive Math Struggle: A 6-Point Action Plan for Fostering Perseverance* and *Answers to Your Biggest Questions About Teaching Elementary Math*. She is also a national mathematics curriculum consultant and speaks at state, regional, and national conferences. She served the National Council of Teachers of Mathematics (NCTM) as the chair of the editorial panel for the journal *Teaching Children Mathematics*, as department editor for *Mathematics Teacher: Learning and Teaching PK–12*, and as a member of program committees for annual meetings and regional conferences. Susie was recently elected to the board of directors for NCSM as Regional Director, Central 2.

PART 1

FIGURING OUT FLUENCY FOUNDATIONS

Key Ideas

WHAT IS FLUENCY IN MATHEMATICS AND WHY IS IT IMPORTANT?

In construction, a foundation is the load-bearing part of a building. So it is with fluency—fluency is built upon foundational concepts and skills. Without such foundations, students are unable to build fluency with basic facts, whole numbers, and more. What makes a strong foundation is based on what one is trying to build. So, we start with the goal of procedural fluency. Try out these problems using any strategy that you like:

$$398 + 535$$

$$504 - 495$$

$$1,435 \div 7$$

How did you find the sum of the first example, the difference in the second, and the quotient in the third? Did you use strategies or algorithms? Did you start with one strategy and shift to another? Each of these problems can be solved efficiently using a strategy other than the standard algorithms. For example, in $1,435 \div 7$, the dividend can be broken apart into $1,400 + 35$, and each part can be divided by 7, resulting in $200 + 5$. A person who demonstrates fluency with division notices the following:

- The dividend (1,435) includes multiples of 7.

  Foundation: Knowing multiples (Module 8)

- The dividend can be decomposed into more noticeable multiples of 7 (1,400 + 35; not decomposed by place value).

  Foundation: Being able to flexibly decompose (Module 2)

- If $14 \div 7 = 2$, then $1,400 \div 7 = 200$.

  Foundation: Multiplying by tens and hundreds (Module 7)

FIGURE 1 ● The Relationship of Different Fluency Terms in Mathematics

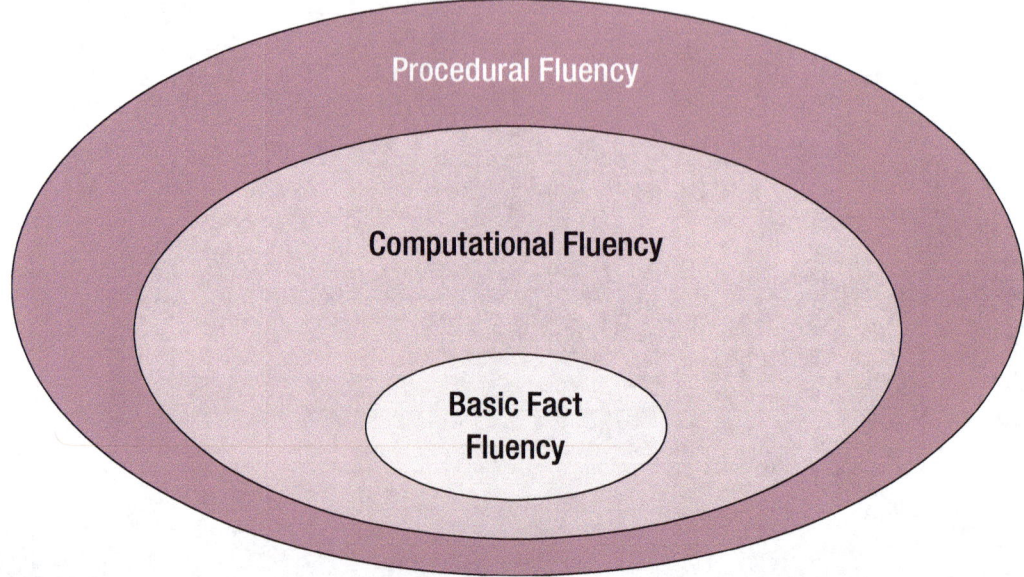

Importantly, with these foundational concepts and skills in place, a person has access to a strategy that is more efficient and less error-prone than long division. Thus, these foundations are the necessary and good beginnings of fluency! Revisit the other problems posed above and ask yourself, "What foundational concepts and/or skills allow me to solve this problem more efficiently than using a standard algorithm?"

Procedural fluency is an umbrella term that includes basic fact fluency and computational fluency (see Figure 1).

Basic fact fluency attends to fluently adding, subtracting, multiplying, and dividing single-digit numbers (see Figure 2).

FIGURE 2 ● Basic Fact Strategies and Their Extensions

BASIC FACT STRATEGY	BASIC FACT (SINGLE DIGIT) EXAMPLE	EXTENSIONS TO OTHER TYPES OF NUMBERS
Making 10	$7 + 9 = 6 + 10 = 16$	$97 + 35 = 100 + 32$ $3.9 + 1.4 = 4 + 1.3$
Pretend-a-10 (Compensation)	$9 + 6 \rightarrow 10 + 6 \rightarrow 16$ $16 - 1 = 15$	$3{,}499 + 5{,}148 \rightarrow 3{,}500 + 5{,}148 - 1$
Think Addition	$11 - 7 \rightarrow 7 + ? = 11$	$89 - 75 \rightarrow 75 + ? = 89$ $9\frac{1}{8} - 8\frac{1}{2} \rightarrow 8\frac{1}{2} + ? = 9\frac{1}{8}$
Doubling	$4 \times 7 = 2 \times 7 \times 2$	$4 \times 2\frac{1}{2} = 2 \times 2\frac{1}{2} \times 2$ $5 \times 28 = 5 \times 2 + 14$
Add-a-Group	$6 \times 7 = 5 \times 7 + 7$	$26 \times 4 = 25 \times 4 \times 4$
Subtract-a-Group	$9 \times 8 = 10 \times 8 - 8$	$99 \times 8 = 100 \times 8 - 8$
Think Multiplication	$45 \div 9 \rightarrow 9 \times ? = 45$	$14.35 \div 7 \rightarrow 7 \times ? = 14.35$

Computational fluency refers to the fluency in four operations across number types (whole numbers, fractions, etc.), regardless of the magnitude of the number. Procedural fluency encompasses both basic fact fluency and computational fluency plus other procedures, such as finding equivalent fractions.

Procedural fluency is defined as solving procedures efficiently, flexibly, and accurately (National Council of Teachers of Mathematics [NCTM], 2014; National Research Council, 2001). The meaning of these three components are

  Efficiency: Solving a procedure in a reasonable amount of time by selecting an appropriate strategy and readily implementing that strategy.

  Flexibility: Knowing multiple procedures and applying or adapting strategies to solve procedural problems (Baroody & Dowker, 2003; Star, 2005).

  Accuracy: Correctly solving a procedure.

To focus on fluency, we need specific, observable actions that we can look for in order to assess what students are doing as they solve computational problems. We have identified six such actions. The three components and six fluency actions (and their relationships) are illustrated in Figure 3.

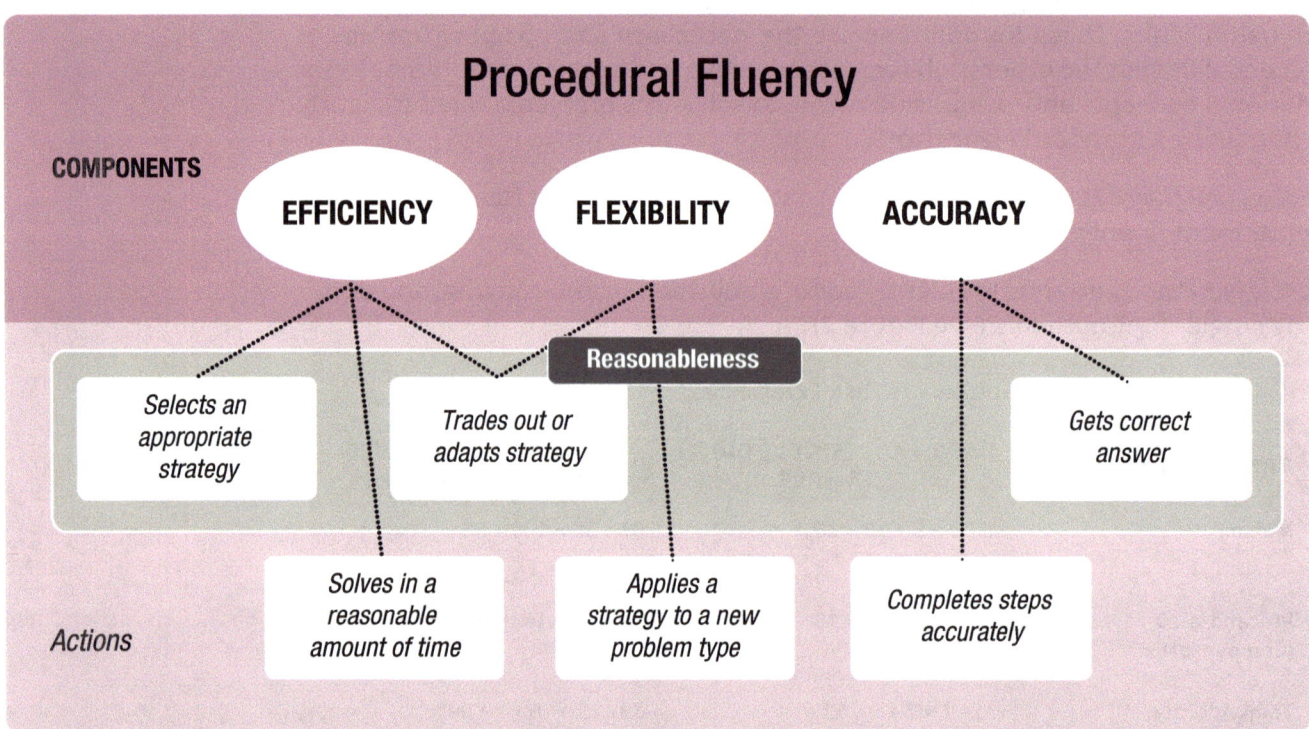

Three of the six fluency actions (should) attend to reasonableness. Fluency actions and reasonableness are described later in Part 1, but first, it is important to consider why this bigger (comprehensive) view of fluency matters. Real fluency is the ability to select efficient strategies; to adapt, modify, or change out strategies; and to find solutions with accuracy.

Real fluency is not the act of replicating someone else's steps or procedures for doing mathematics. It is the act of thinking, reasoning, and doing mathematics on one's own. The NCTM (2023) Procedural Fluency Position Statement describes what procedural fluency is and what is necessary to ensure all students develop procedural fluency, citing significant research along with instructional resources for classroom support.

TEACHING TAKEAWAY

Real fluency is the ability to select efficient strategies; to adapt, modify, or change out strategies; and to find solutions with accuracy.

WHAT DO FLUENCY ACTIONS LOOK LIKE FOR THE OPERATIONS?

The six fluency actions are observable and therefore provide insights into the foundational knowledge and skills students require. Each one is briefly described here, connected to the problems posed at the start of this section.

TEACHING TAKEAWAY

Selecting *an* appropriate strategy does not mean selecting *the* appropriate strategy. Many problems can be solved efficiently in more than one way.

FLUENCY ACTION 1: Select an Appropriate Strategy

Selecting *an* appropriate strategy does not mean selecting *the* appropriate strategy. Many problems can be solved efficiently in more than one way.

Here is our operational definition:

> **Of the available strategies, the one the student opts to use gets to a solution in about as many steps and/or about as much time as other appropriate options.**

Consider 398 + 535. A student might start with 398, jumping up 500, then 2, and then 33 more (Count On strategy, see Figure 4). Another student might move 2 from 535 to 398 to make 400 and solve it (Make Hundreds strategy, see Figure 4). Or a student might leave 535 alone, add 400, and subtract 2 from their answer (Compensation strategy, see Figure 4). Each of these are appropriate for this problem because they each take about as many steps as the others. The standard algorithm, however, is not an appropriate choice, given the additional steps and time it would take to enact these addends.

FIGURE 4 ● Reasoning Strategies for Adding 398 + 535

COUNT ON	MAKE HUNDREDS	COMPENSATION
	$398 + 535$ $\wedge$ $2\ \ 533$ $398 + 2 = 400$ $400 + 533 = 933$	$398 + 535$ $+2$ $400 + 535 = 935$ $935 - 2 = 933$

Important points about these strategies include the following:

- They may be mental or written.

- They are flexible (there are other ways to use Count On, for example).

- The choice of a strategy requires fluency foundations—noticing that 398 is 2 away from 400, in this case (Make Hundreds strategy).

- The enactment of a strategy requires fluency foundations—decomposing and skip counting, for example, could be utilized for the Count On strategy.

FLUENCY ACTION 2: Solve in a Reasonable Amount of Time

The time it takes to solve a problem depends on the numbers in the problem and the mathematical maturity of the solver. A reasonable amount of time attends to two things: (1) the enactment of the selected strategy is efficient (e.g., Counting On in chunks rather than Counting On by ones) and (2) the solver works through their strategy without getting stuck or lost. For example, a student solving 504 – 495 may have noticed that they could use Think Addition (Count Up) and then drew a number line and counted by ones from 495 up to 504. This would be reasonable for a younger student learning subtraction as "find the difference," but with maturity, this strategy would be quicker, likely done mentally and by chunking the jumps (+5 to 500 and + 4 to 504).

FLUENCY ACTION 3: Trade Out or Adapt a Strategy

As strategies are better understood, students are able to adapt them or swap them out for another, more efficient strategy. For example, a student solving 1,435 ÷ 7 may first attempt to break apart the dividend by place value and get stuck because 1,000 is not a multiple of 7. Then, they decide to use Think Multiplication, reasoning that there are 100 sevens in 700—so 200 sevens in 1,400—and 5 sevens in 35, which add up to 205. When a strategy is not going well, a student goes back to other options, looking at the problem to see what might work. Similar to the original selection of a strategy, this is when a person relies on foundational understandings and skills to choose and enact a strategy.

FLUENCY ACTION 4: Apply a Strategy to a New Problem Type

Take a strategy like compensation. It can be used with basic facts (e.g., thinking of 9 + 7 as 10 + 7 and take away 1), whole numbers (see Figure 4 above), and with fractions or decimals. Students generalize the idea that they can adjust a problem to make it easier to compute, and then they compensate to preserve equivalencies. Such generalizations are the properties in action!

FLUENCY ACTIONS 5 AND 6: Complete Steps Accurately and Get Correct Answers

An error at the end of a problem may be due to an error in how a strategy was enacted or due to an incidental error. For example, a student may think of the Halve and Double strategy accurately to solve $4 \times 3\frac{1}{2}$, as illustrated in Figure 5 by halving 4 and doubling $3\frac{1}{2}$, but make a computational error (doubling 4 instead of halving it).

FIGURE 5 ● Halve and Double Strategy Is Implemented Correctly, but a Computational Error Is Made

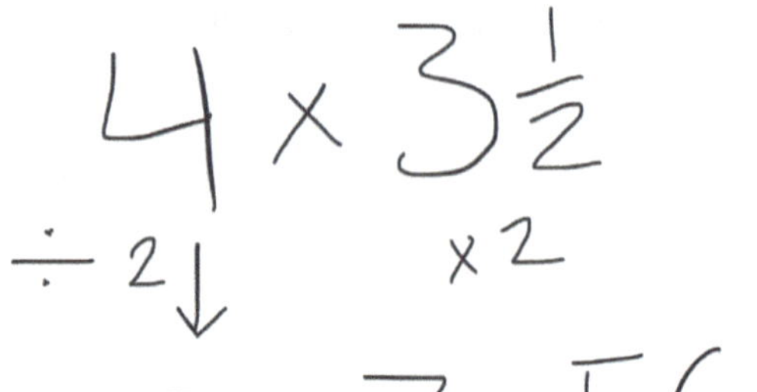

$$4 \times 3\tfrac{1}{2}$$
$$\div 2 \downarrow \qquad \times 2$$
$$8 \times 7 = 56$$

As these fluency actions indicate, true fluency requires decision-making, and those decisions require foundational understandings and essential skills. Procedural fluency is important for life and for higher-level mathematics. Most importantly, unrealized fluency creates significant barriers to students' productive and positive mathematics identity and agency. It all begins with ensuring students develop strong foundational understandings and skills!

REASONABLENESS

Reasonableness is more than checking your answer; it occurs in three of the six fluency actions as shown in Figure 3. Let's explore reasonableness for the problem 504 – 495. These two numbers are close together, thus Think Addition (Counting Up) is a reasonable strategy choice (Action 1). Carrying it out "reasonably" means to monitor if the selected strategy is going well (e.g., Count Up by Ones) and if not, to adapt it (e.g., count up by 5 to get to 500 and + 4 to get to 504) or trade out the strategy (Action 3). Finally, 9 is a reasonable answer because the numbers are close together (Action 6). At each of these phases, we see the role of foundations:

- To start, one must notice the relative size of the numbers.
- To chunk is to know the distance to 100.
- To know the answer is reasonable, one can use computational estimation (e.g., distance from 500).

Reasonableness is essential for fluency. The three Cs of Reasonableness (Choose, Change, Check) can provide strong support for students as they are thinking through a problem (see Figure 6). Importantly, a focus on reasonableness also supports the development of foundations and vice versa. For example, in asking, "Is this something I can do in my head?" a student will take time to look at the numbers in the problem and look for multiples or proximity to a benchmark, and the more they look for these relationships, the better they get at multiples or determining the distance to a benchmark.

FIGURE 6 ● Choose, Change, Check Reflection Card for Students

Checks for Reasonableness		
Choose	Change	Check
Is this something I can do in my head? What strategy makes sense for these numbers?	Is my strategy going well, or should I try a different approach? Does my answer so far seem reasonable?	Is my answer close to what I anticipated it might be? How might I check my answer?

Icon sources: Choose by iStock.com/Enis Aksoy; Change by iStock.com/Sigit Mulyo Utomo; Check by iStock.com/Indigo Diamond.

THE TEN FOUNDATIONS

Good and necessary beginnings—foundations—include the concepts and skills that are essential to reasoning. The example in the preface of 8 + 5 illustrates the key role of foundational understandings and skills to enact the Making 10 strategy. As another example, consider this multiplication problem: 49 × 15. Here is one way to solve this problem (which could be done mentally or in writing):

$$49 \times 10 = 490$$

$$49 \times 5 \text{ is half of } 490 \text{ (so } 245)$$

$$49 \times 15 = 490 + 245$$

$$= 500 + 235$$

$$= 735$$

Within this process, a student needs to understand the distributive property, be able to multiply by tens, find half of a number, and decompose to add. This one example clearly illustrates the critical need to ensure students have foundational knowledge and skills. So what are those good and necessary beginnings? Figure 7 provides an at-a-glance list of ten foundations that are necessary in giving students access to doing mathematics.

FIGURE 7 ● Foundations for Computational Fluency

FOUNDATIONS MODULES	WHAT IT IS (IN BRIEF)	EXAMPLE PROMPT
1. Number Relationships: Comparison and Estimation	Knowing approximately where a number is in relation to other numbers	Please place 28 on a number line. Use various endpoints [0, 50], [20, 30], or [0, 100].
2. Subitizing and Decomposing	Subitizing is visually seeing subsets of the whole to determine the whole; decomposing is starting with the whole and determining the parts	 How many dots? How do you see them? Decomposing: There are 10 turtles on two logs. How many might be on each log?
3. Distance to 10, 100, and 1,000	Seeing how far a number is from a benchmark (over or under)	How far is … ● 8 from 10? ● 37 from 40? ● 107 from 100? ● 288 from 300?
4. Counting and Skip Counting	Starting with a number and counting on or back by ones or other intervals (e.g., 20, 25)	Start at 48. Skip count by tens (or twenties). Start at 337 and count back 80.
5. Properties of Addition and Its Inverse Relationship With Subtraction	Using the commutative and associative properties of addition and using the relationship between addition and subtraction (i.e., if $a + b = c$, then $c - a = b$)	Give students an equation (for example, 25 + 15 = 40) and ask, "What other equations are true, using these same numbers?"
6. Properties of Multiplication and Its Inverse Relationship With Division	Using the commutative and associative properties of multiplication, the distributive property of addition over multiplication, and the relationship between multiplication and division (i.e., if $a \times b = c$, then $c \div a = b$)	Ask students, "Does 6 × 4 have the same answer as 4 × 6?" Then ask students to show/explain how they know it is true.
7. Multiplying by Tens and Hundreds	Being able to multiply any number (e.g., 16, 135, 5.2) by a multiple of 10 and know why it works Similarly understanding that 60 × 9 is the same as 6 × 9 × 10	Ask students to multiply 60 × 9. Ask, "How did you think about that?" Listen for answers that indicate understanding, not rule-based, incorrect explanations such as "I added a 0 on 54." Ask students to show 2,400 ÷ 6.

FOUNDATIONS MODULES	WHAT IT IS (IN BRIEF)	EXAMPLE PROMPT
8. Multiples and Factors	Recognizing when a basic fact is present (such as in $2{,}400 \div 6$) and using that relationship to solve the problem	Give students two numbers (for example, 12 and 18) and ask what is alike and different about these numbers. If they don't focus on multiples and factors, then prompt for such responses.
9. Doubling and Halving	Being able to readily double or halve numbers (e.g., 48 or 250)	Ask students to double 36. If they are stuck, ask to double 30, then 6, and return to 36. Ask students to halve numbers with even digits (e.g., 264) and odd digits (e.g., 634).
10. Computational Estimation	Being able to quickly determine an answer close to the actual answer by using an estimation strategy (which does not include finding the exact answer and rounding it!)	Ask, "About how much is the answer?": $57 + 68$ $402 - 189$ 19×9 $253 \div 6$ Ask how they thought about it.

Each of these foundations are a full module in this book, which can be taught as a unit (replacing what might be in place for that foundation), used as a supplement (textbook coverage is often not sufficient to develop deep understanding and automaticity with these foundations), or used for interventions (because students who struggle with computation are often in need of more support with a foundational concept or skill).

For intervention, the use of these modules begins with figuring out which foundations are priority for the student. You have several options (using the table above) to decide where a student's strengths and needs are.

1. You can use the questions in Figure 7 to get a feel for what the student can do. Once you notice an area of need, stop there or move to a question they are likely to know well. It is counterproductive to go through a series of prompts wherein the student is struggling and experiencing stress or anxiety.

2. Go to the modules and read the assessment section at the beginning. There, you will find what you really need to be looking and listening for, along with six quick assessments (prompts) that lend to gaining insights into the students' understanding and skill.

3. Determine if the student needs to better understand the foundation conceptually. The first five activities in each module lean toward instruction for understanding the foundation. The remaining activities in a module provide opportunities for repetition. These are designed for students who show understanding but need more opportunities to work with the skill so that it becomes automatic and usable.

Part 3 helps you think about implementation. A series of frequently asked questions (FAQs) are provided for implementation in various settings—the classroom, the intervention setting, and at home. In addition, more ideas are provided for monitoring student success and ensuring that we continue to attend to students' emerging mathematics identities and agency as we engage in developing strong foundations within initial classroom instruction or intervention.

PRODUCTIVE BELIEFS ABOUT FLUENCY AND ITS FOUNDATIONS

With fluency defined, it is important to state that every student can develop procedural fluency. Attaining fluency for every student requires productive beliefs about fluency, described in Figure 8, which is also in our *Figuring Out Fluency* anchor book.

FIGURE 8 ● Productive Beliefs About Procedural Fluency

1. Procedural fluency is an attainable goal for each and every student. Each student is capable of developing a repertoire of strategies and learning skills at applying those strategies flexibly, efficiently, and accurately.

2. Procedural fluency is a function of opportunity, experience, and effort. Differentiated supports enable each and every student to understand and use a range of strategies.

3. Procedural fluency instruction is higher-order thinking, as students create strategies, generalize when to use a strategy, and explain why a strategy works. This increased level of thinking leads to greater understanding and performance for every student.

4. Every student must have access to instruction and resources that attend to all procedural fluency components and actions.

5. Having a range of ideas and strategies for solving procedures enriches everyone's learning. Therefore, every student benefits from heterogenous grouping; conversely, homogeneous grouping (ability grouping) is detrimental to developing procedural fluency.

Productive Belief #1 is easy to agree with but not easy to enact. What does it look like to engage students in ways that say to them, "You are capable. You can figure this out"? A start is to shift *away from* showing students how to do something and move *toward* students showing us how they did what they did. This shift requires more of a teacher, not less. This leads into Productive Belief #2: providing opportunities and experiences to support students' development of procedural fluency. Students need quality and substantial foundations in order to eventually develop fluency with an operation or procedure. In our *Figuring Out Fluency* anchor book, we describe these as Good (and Necessary) Beginnings for Fluency (Chapter 3). In that chapter, we describe conceptual understandings, properties, utilities, and skills that enable students to reason with basic facts and beyond. The *Figuring Out Fluency* companion books (see list in the preface) focus on developing fluency (e.g., with whole-number addition and subtraction), yet these books could not fully take on the readiness skills to set students up for success—that is the purpose of this book! *Figuring Out Fluency—Ten Foundations for Reasoning Strategies With Whole Numbers* lands squarely on the foundations students need in order to have success with the strategies developed in the other books.

HOW DOES CONCEPTUAL UNDERSTANDING DEVELOP STRONG FOUNDATIONS (AND FLUENCY)?

Conceptual understanding is connected knowledge: "mental connections among mathematical facts, procedures, and ideas" (Hiebert & Grouws, 2007, p. 380). For a subtraction problem, for example, conceptual understanding includes knowing the relative size of the numbers and understanding that subtraction can be interpreted as "find the difference" or "take away," that there are various strategies for finding answers to subtraction problems, and the various ways to represent those problems with manipulatives or drawings.

The NCTM offers this declaration related to the relationship between concepts and procedures: "Conceptual understanding must precede and coincide with instruction on procedures" (2023, p. 2). The elaboration explains that learning is supported when instruction on procedures and concepts is explicitly connected and iterative. Conceptual foundations lead to opportunities to develop reasoning strategies, which in turn deepens conceptual understanding. This is consistent with the classic concrete—semi-concrete—abstract (CSA) sequence (Bruner & Kennedy, 1965; Flores et al., 2018; Griffin et al., 2014). The CSA model is not a linear progression either. By design, the intent is to loop back to the C and the S to make sense of the A. Here, we highlight important ways to help students make connections and thereby develop strong foundations for fluency.

TOOLS AND REPRESENTATIONS

Manipulatives and visuals make mathematical relationships visible so that students can internalize abstract concepts. For example, the representations in Figure 9 help students see the relationship among numbers as well as the relative size of a number.

FIGURE 9 ● Representations to Show the Relative Size of the Number 8

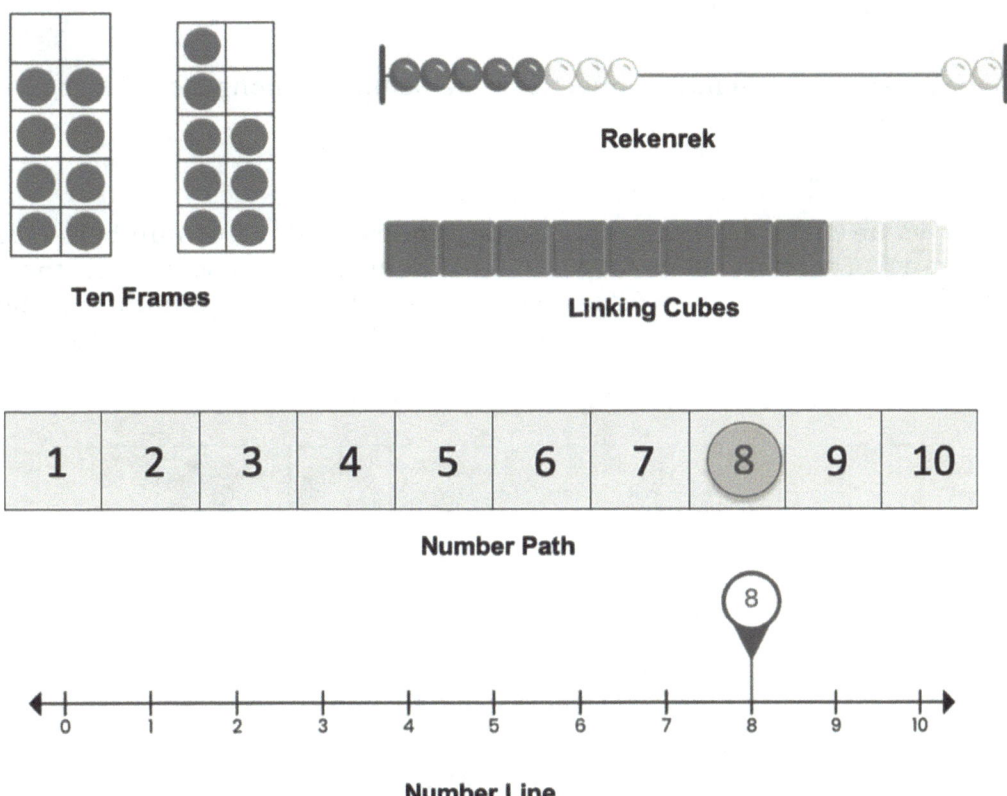

Ten Frames

Rekenrek

Linking Cubes

Number Path

Number Line

There is no shortage of objects that can be used to count. These objects can be used for subitizing, decomposing, exploring part–part–whole, and much more. Many tools, such as the ten frame, can initially be concrete (wherein students physically move counters on and off ten frames to illustrate numbers) then become visuals (similar to that pictured in Figure 10). And they may become mental images to support abstract reasoning. Here, we share four commonly used representations, offering some insights about the tool and how to use it.

TEN FRAMES

Ten frames can be used for subitizing, decomposing, number combinations, and implementing reasoning strategies such as Make Tens. A full row can be filled first to illustrate a number's relationship to 5, but it can be filled any way you choose to highlight a number relationship. Ten frames can be presented vertically or horizontally (as illustrated in Figure 10). A vertical orientation aligns with expressions written horizontally and, conversely, a horizontal orientation fits with a problem that is stacked.

FIGURE 10 ● Representing Addition on Ten Frames

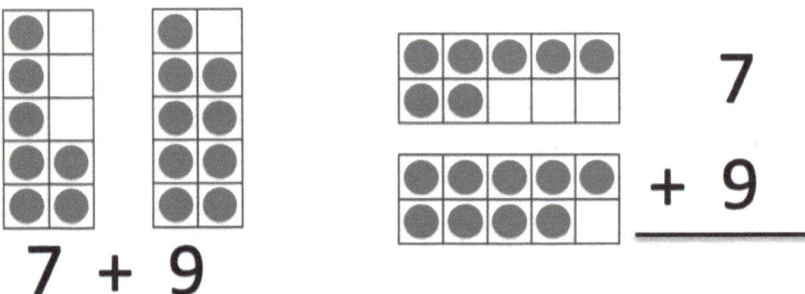

In both cases, we see how connections can be made among facts, ideas, and procedures.

PLACE-VALUE DISKS WITH TEN FRAMES

Place-value disks are nonproportional representations of numbers (see Figure 11). There are disks for ones, tens, hundreds, and so on. Pair them with ten frames for powerful representations of multi-digit numbers that can help students see how ones, tens, and hundreds can be regrouped.

FIGURE 11 ● Place-Value Disks and Ten Frames Showing 385

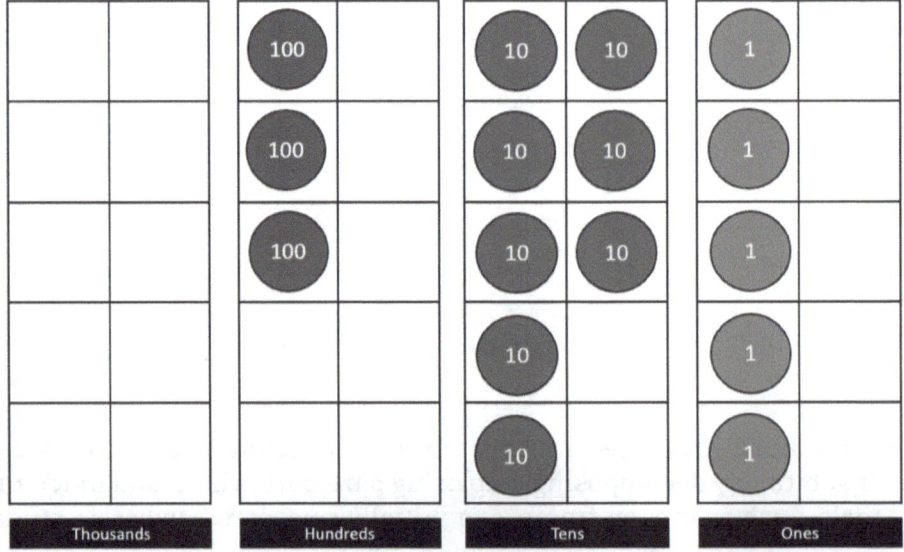

NUMBER PATHS AND NUMBER LINES

There is strong evidence that using a number line facilitates learning concepts and procedures with grade-level content and future learning (Fuchs et al., 2021). Being able to place numbers on a number line predicts student success years later (Geary, 2011). Number paths help students see quantity while also seeing how far numbers are from 0 or 10. They serve as an excellent tool for helping make sense of the abstract number line. Number paths and lines can also be positioned vertically and horizontally. In fact, if students are representing a situation that is vertical, then a vertical number line makes more sense. For example, if students are comparing heights of cubes or plants, then a vertical number line makes sense. Open number lines are particularly useful in reasoning because students do not need to get bogged down in the accuracy of unit lengths but they rather approximate the lengths to illustrate their thinking.

BOTTOM-UP HUNDRED CHART

The hundred chart is commonplace in Grade 1, 2, and 3 classrooms, yet its classic orientation is upside down (Bay-Williams & Fletcher, 2017). Figure 12 displays the Bottom-Up Hundred Chart. In this position, values go *up* on the chart as they *increase* quantitatively. This idea can and should be extended to decimal charts as well. Not only is this a stronger connection to the operations, but it is also more like the coordinate axis. A hundred chart helps students see place-value concepts and develop relational understanding (seeing that 48 is close to 50 and 10 away from 58). A hundred chart can also be cut in rows and taped together to form a 100 number path, which is an excellent bridge to working with number lines.

FIGURE 12 ● Bottom-Up Hundred Chart

91	92	93	94	95	96	97	98	99	100
81	82	83	84	85	86	87	88	89	90
71	72	73	74	75	76	77	78	79	80
61	62	63	64	65	66	67	68	69	70
51	52	53	54	55	56	57	58	59	60
41	42	43	44	45	46	47	48	49	50
31	32	33	34	35	36	37	38	39	40
21	22	23	24	25	26	27	28	29	30
11	12	13	14	15	16	17	18	19	20
1	2	3	4	5	6	7	8	9	10

 This resource can be downloaded at **https://qrs.ly/psf6a5o**

PARTS AND WHOLE RECTANGLE

Addition is the joining of parts to make a whole, and subtraction is seeing the difference between a whole and a part. Thus, the part–part–whole graphic (Figure 13) provides a layout that illustrates this relationship, helping students see quantitative relationships, and can be used to place manipulatives such as linking cubes (see Activity 2.2), counters, or base ten pieces. Students can be given the whole and asked to find one or both parts or be given parts and asked to find the whole.

FIGURE 13 ● Part–Part–Whole Placemat

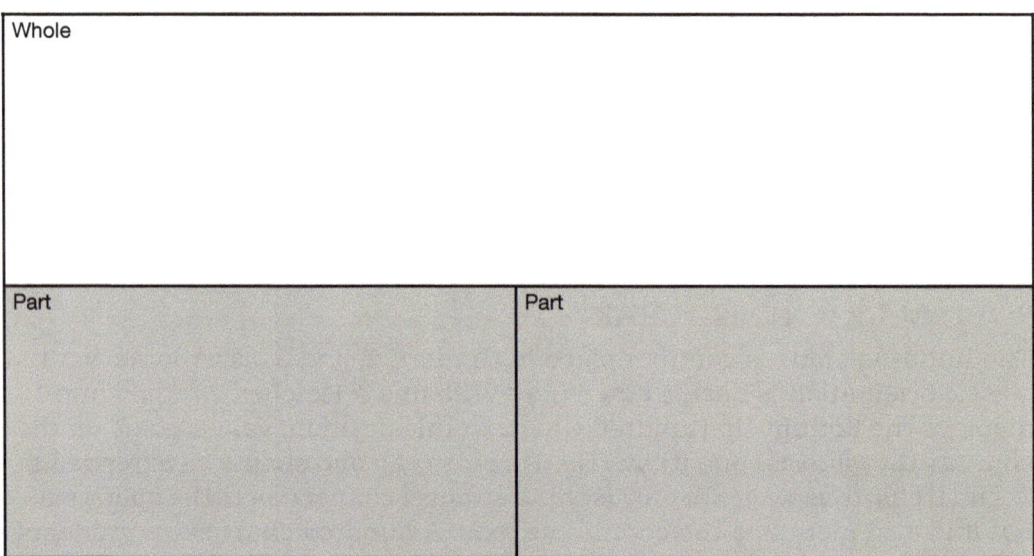

online resources ⬉ This resource can be downloaded at **https://qrs.ly/psf6a5o**

Eventually, students can replace objects with numbers: for example, placing numbers from a story problem on separate sticky notes and deciding where to place them to represent the story. This is true with whole numbers, fractions, or decimals. While many parts and whole visuals have two parts, they could have three or more parts, which simply involves adapting the part–part–whole placemat template (which is available in the downloadable slides). It can also be positioned vertically or horizontally.

MATHEMATICAL LANGUAGE

Similar to number lines, there is strong evidence that supporting students' use of mathematical language will support their learning of mathematics (Fuchs et al., 2021). Mathematics is based on very precise language, where meanings of words can change in different circumstances (e.g., *increased by* and *multiply by*) and where one word can make a difference in what operation is needed (e.g., *How many?* versus *How many more?* or *How many more?* versus *How many times more?*). As we listen to students describe their foundational concepts or skills, we need our own skills to help students develop this precision. Asking students to restate or rephrase can help teachers assess understanding while helping students learn the language of mathematics (Chapin et al., 2013). Example prompts include the following:

- "So, you said ..." [revoice, inserting more precise language]

- "You used the hundred chart and counted on ...?" [paraphrase using mathematical language]

- "Please repeat what you/someone just said." [listen for precise language and understanding]
- "Explain ____ using the words _____ and ____ in your explanation." (e.g., "How might you explain what you did with the linking cubes using the word *decompose?*")

It's also powerful to have students revoice their thinking while they practice and as they play games with partners, as it helps them

- clarify their process by hearing their thinking,
- strengthen metacognition and retention,
- provide another practice exposure to their partner, and
- gain insight into someone else's strategy or process.

A second element to mathematical language is the broader goal of getting students to talk (classroom discourse). The moves described above support this goal as well. Think-alouds and peer tutoring are consistently found to positively impact student learning. And when students are talking about their thinking, teachers are getting a much stronger sense of what the student understands. Think—pair—share is an age-old yet underutilized classroom practice that ensures all students have processing time, are expected to articulate their thinking, and have the opportunity to learn from others.

WHAT DOES QUALITY PRACTICE LOOK LIKE FOR THE FOUNDATIONS?

Fluency practice is not a worksheet! This is the title of Chapter 6 in our anchor book, *Figuring Out Fluency in Mathematics Teaching and Learning*. Worksheets do not support good beginnings! Figure 14 provides a visual to capture the elements of quality practice, with example tasks within each.

FIGURE 14 ● Quality Practice Is Not a Worksheet!

Quality Practice Is . . .

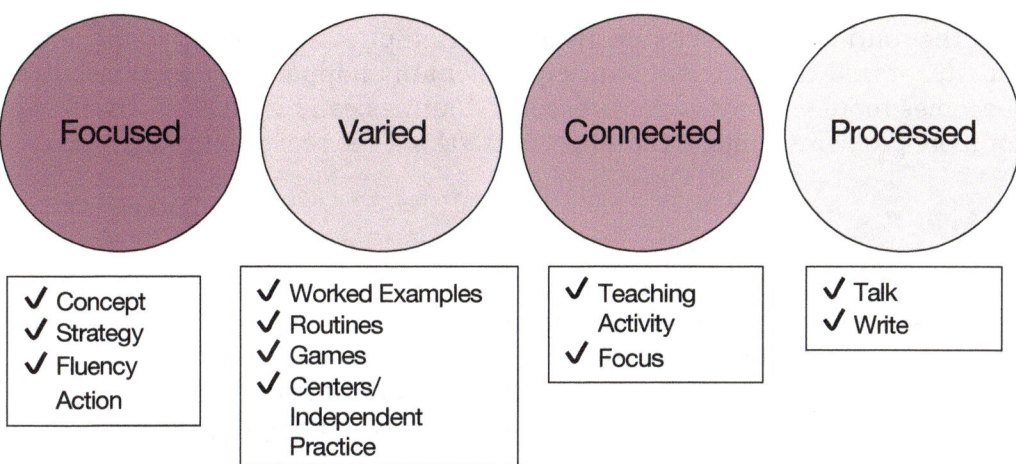

Focused
- ✔ Concept
- ✔ Strategy
- ✔ Fluency Action

Varied
- ✔ Worked Examples
- ✔ Routines
- ✔ Games
- ✔ Centers/Independent Practice

Connected
- ✔ Teaching Activity
- ✔ Focus

Processed
- ✔ Talk
- ✔ Write

This is not to say that practice requires a lot of time—quite the contrary. Short but ongoing, engaging practice is what is needed. If practice is too long, it can become unproductive. If it is too short, students don't get enough practice to solidify their understanding or develop the automaticity they need with the skills. The right amount of time gives enough exposure and keeps students engaged. There is no set number of minutes. It varies from practice activity to student to topic.

ROUTINES

A routine is a familiar, adaptable protocol for engaging students in learning through thinking and discussion. Many routines also use representations. Thus, routines help students build conceptual understanding and make connections to strengthen their emerging foundational concepts and skills. Routines can foster positive mathematics relationships within the classroom community (Berry, 2018). The exchange of ideas during a routine is essential for advancing student understanding and fluency. Discussion within routines reassures students that their emerging, possibly less common strategies are viable and used by others. Learners build confidence when they see that their strategy is reasonable and taken up by others.

Every module has several routines that focus on the selected foundation. But these routines are often adaptable to other foundations or to reasoning strategies. The keys to using routines effectively are to do the following:

1. Ensure all students understand the purpose of the routine and practice the steps of the routine.

2. Allow individual think time.

3. Make space for partner work.

4. Conduct full-group discussions of ideas.

5. Pose questions that focus on the key mathematical goal of the routine.

6. Encourage multiple representations and ideas.

7. Resist injecting your ideas or approaches too soon (or at all).

8. Keep routines short (5 to 10 minutes).

Try the routines but don't stop after one attempt. It is typically in the fourth or fifth round that students anticipate what is happening and the routine becomes more productive for everyone. Routines can be adapted—use fewer or more problems, change the steps, or trade out the representation.

GAMES

Games offer enjoyable practice. That joy is a benefit, not a rationale. The reason games are quality practice is because, similar to routines, they engage students in talking about their reasoning and listening to each other's strategies and provide an opportunity for teachers to listen and formatively assess. If you were to tally all the problems a student solves when playing a game, the list would certainly fill a worksheet. So, games provide opportunities for substantial and meaningful practice ... at least, they *should*. Games that have a time component rob students of their processing time, add stress, and communicate that being good at mathematics means being fast. Games should not be timed nor should they pit students against each other in attempting to solve the same problem most quickly.

TEACHING TAKEAWAY

Games should not be timed and should not have students solving the same problem.

Because games are fun, students can forget they are practicing a mathematical skill. Tips to maximize learning when playing games include the following:

1. Tell students the purpose of the game and have them tell it back to you.

2. Give students sentence frames to help them articulate their thinking.

3. Provide recording sheets for students to record the problems they encountered.

4. Have students revoice their thinking as they complete a turn.

5. At the end of the game, ask students to reflect on their growth related to the skill they were practicing as well as any insights they gained about the mathematical ideas.

Games, similar to routines, can always be adapted. You can adapt games yourself or ask students how they would like to adapt the game. Some adaptations can simplify or increase the mathematical challenge; others can increase/decrease the complexity of the game itself (which will vary based on the age of students).

CENTERS

Centers are physical locations in the classroom set up with a mathematics activity that students can explore independently (alone or with a partner). Centers have traditionally been reserved for younger grades but are appropriate for all grades. Centers may have sorting tasks, choice problems, or games that can be played independently. Where routines and games provide opportunities for students to use language and learn with their peers, center activities provide extended time for individual engagement with a concept or skill. As students engage with the activity, they complete a recording page, providing themselves and you with a written record of their reasoning. Center activities can also be sent home or used as classroom activities. Students can work with partners or alone.

WHAT ARE THE RELATIONSHIPS AMONG TEACHING, PRACTICING, AND ASSESSING?

In Part 1, we have so far elaborated on what fluency means and thus what foundations become critical for students—in other words, the necessary and good beginnings. Throughout that discussion, representations and opportunities for students to use mathematical language took front and center stage. Where do representations and language use fit into the teaching, practicing, and assessing aspects of teaching? Everywhere! In fact, there is a lot of overlap and multi-purposing across teaching, practicing, *and* assessing activities. Students may be given a quick assessment for you to know where to focus instruction, but for them, it is also an opportunity to learn and to practice.

As you engage students in a game and require that they think aloud, you can assess their thinking as you manage the activity. The visual in Figure 15 illustrates that we use a variety of representations, visuals, and activities within each of these domains.

FIGURE 15 ● Ways to Teach, Practice, and Assess to Support Foundations and Fluency

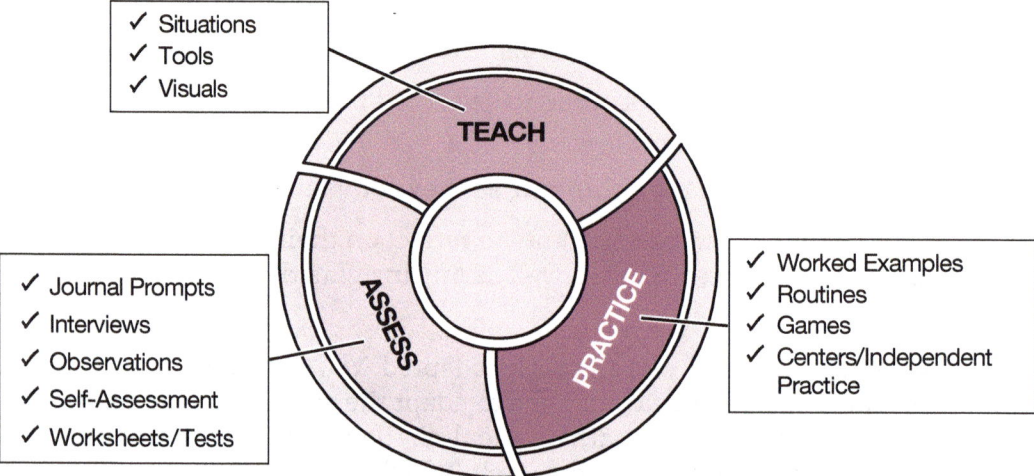

This graphic can help you continue to vary the way in which students learn, practice, and are assessed. A list of all of the teaching and practice activities are provided in the Appendix. Keep your eye on the big ideas discussed in Part 1, and then flip to Part 3 for FAQs and ideas for implementing the modules in the classroom, during intervention, and in other settings.

PART 2

TEN FOUNDATIONS
FOR FLUENCY

Number Relationships: Comparison and Estimation

NUMBER RELATIONSHIPS OVERVIEW

In mathematics, students must understand and compare quantities. Comparing quantities involves understanding the relative size of a number; for example, knowing that 8 is 1 more than 7, which also means 7 + 1 = 8 and 8 − 1 = 7. This grows into noticing that 47 is 3 away from 50 and 6.92 is 8 hundredths away from 7. Importantly, comparing is not about using the < and > symbols correctly; rather it is knowing that 8 is more than 7 because it is one more object and it is the number to the right on the number line. Estimation is similar to comparison but allows for approximation: for example, noticing that 47 is close to (but to the left of) 50. Students need to be able to find the approximate location of a number on a number line as it relates to benchmarks. Understanding these two number relationships develops students' number sense. Together, they lay the foundation for estimating sums and differences, which is the focus of Module 10.

In this module, the focus is on placing numbers on number lines, changing their locations as endpoints change, and reasoning about how close numbers are to one another. For many students, this work is often procedural. Comparing numbers is consigned to looking at place value. Students may have limited work with number lines, and the experiences they do have may imply that number lines are static—always having the same endpoints (e.g., 0 and 100). Connecting estimation to comparison and number line work is challenging because estimation doesn't typically play a large role in mathematics classrooms. Also, students can misinterpret what it means to estimate, thinking that there is a correct estimation. Perceiving one right way to estimate leads to inflexible, procedural approaches to their reasoning.

HOW DO NUMBER RELATIONSHIPS CONTRIBUTE TO FLUENCY STRATEGIES?

Understanding numbers and their relationships is inseparable from computational fluency. It helps students select a strategy, execute it, and determine if their results are reasonable. In 48 + 25, a student recognizes that 48 is close to 50 and they can use the Make Tens strategy. They break 25 into 2 and 23, giving 2 to 48 to make 50. From there, they add 23, finding the sum of 73.

$$48 + 25$$
$$2 + 23$$
$$48 + 2 + 23$$
$$50 + 23 = 73$$

This also aids in determining reasonableness. This same student can reason that 48 is close to 50 and that 25 is close to 30. They determine that 50 + 30 = 80 so then the sum of 48 + 25 should also be close to 80.

A strong understanding of number relationships also helps students represent their thinking, which is especially useful as numbers become larger or more complex. In both examples, the students manipulate their number lines for convenience. Neither number line uses endpoints of 0 and 1,000.

Students choose good starting places; they know how the numbers change as you move to the right or left. They can count in chunks (Module 5).

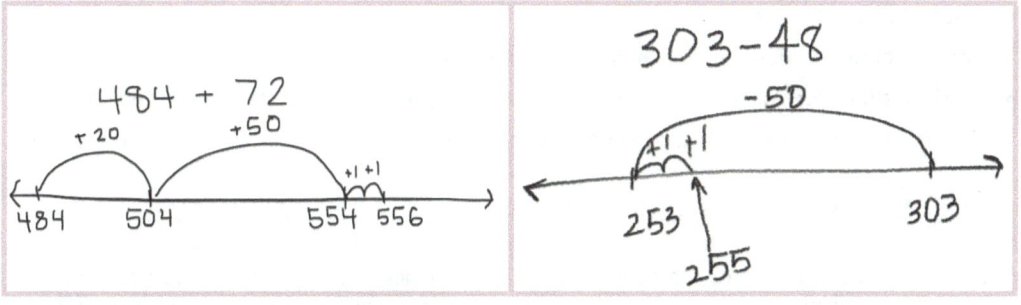

The number relationships reference page provides an overview, important representations, connections to fluency, and actions to avoid. It can be downloaded and used for reference in planning, teaching, and discussions with colleagues and families.

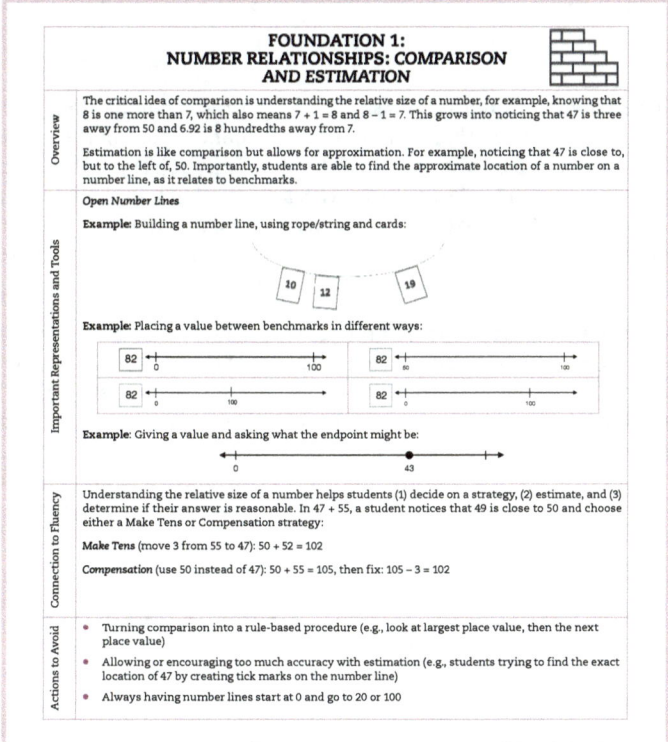

online resources Available for download at **https://online qrs.ly/psf6a5o**

ASSESSING NUMBER RELATIONSHIPS

The goal of assessing student thinking related to number relationships is to see if students have a relational understanding of numbers. Here, we share observation ideas (*look-fors*) and interview or journal prompts (*quick assessments*).

NUMBER RELATIONSHIP LOOK-FORS

Student estimates should improve as they gain more experience estimating. When you ask students to tell you a number that is close to another, pay attention to the numbers they share. Consider whether they always tell you of a number that is 1 or 2 away from a given number. When they work with number lines, be sure that they are reasoning about relationships. They don't have to be exact with their placement, but you do want numbers placed reasonably. For example, 28 should be closer to 0 than 100, 93 should be placed pretty close to 100, and so on. You want to be sure that your students can

- estimate quantities reasonably,
- identify or place a number reasonably on a number line,
- tell how a number's position changes on a number line as the endpoints change, and
- identify numbers that are close to one another and numbers that aren't close to one another.

QUICK ASSESSMENTS FOR NUMBER RELATIONSHIPS

The following tasks are good opportunities to gather evidence of your students' progress. The prompts in the following table, along with the other routines, games, and centers in the practice activities section of this module, are also good options. You can do these in small groups or in a one-on-one setting. They should not take more than a few minutes. You have options, but don't feel obligated to use them all. These examples show two-digit numbers. You can modify the task for three- and four-digit numbers as needed. These examples would also work well with decimals (replacing 56 with 5.6 or 0.56).

Show a group of items (paper clips, buttons, centimeter cubes, etc.). Have students estimate how many they see. Have them explain how they made their estimate. Alternatively, you can give them a collection and ask if a number is a reasonable estimate.	Share a number, such as 56. Ask students to tell you three numbers that are close to 56 and three numbers that are far from 56. Repeat this with a different number. Listen for patterns in their responses.	Draw a number line. Record a benchmark on the number line (e.g., 50). Have students place a few numbers on the number line. Repeat the activity, but this time, use a different benchmark (such as 20).
Ask how a number can be close to more than one other number. Have students give examples to justify their thinking.	Ask students to put different numbers (e.g., 13, 35, 78) on a number line. Then, have them tell you how they knew where to place them. Ask them where they might place another number.	Draw a number line with endpoints of 0 and 100. Ask students where to place a number (e.g., 74). Then, change the endpoints to 50 and 100. Have them place their number again. Then, change the endpoints to 70 and 80. Have them place it again.

EXPLICIT INSTRUCTION FOR NUMBER RELATIONSHIPS OF COMPARISON AND ESTIMATION

Explicitly teaching estimation can be challenging, as you can lead students to think there is always one right answer or way of thinking about numbers. As you teach them to think flexibly, you have to explicitly think aloud often. You have to say things such as, "78 is closer to 100 than 0. It is close to 70 but it's also close to 80." With three-digit numbers, you would say, "612 is closer to 1,000 than 0." You might say, "612 is close to 600, but it's also close to 500." As you likely assume, the idea transfers to four-digit numbers, decimals, and fractions.

You want to help students understand how you estimate but also understand that your approach is flexible. Share the questions you ask yourself as you process estimating and number relationships. You might have to be a bit more flexible yourself, meaning placement of a number on a number line might be off by a little. On a number line with endpoints of 0 and 100, benchmarks of 25, 50, and 75 should be rather precise. However, 38 and 27 should be approximate.

ACTIVITY 1.1
ESTIMATION STATION

Your students can't estimate enough! This is true regardless of how fluent they are with computations, how well they solve problems, or how quickly they learn new content. Estimation is an important life skill. For this activity, you want to gather plastic bags, jars, small baskets, or any other container. Put a random number of objects in each container. You might use two-colored counters, teddy bear counters, centimeter cubes, or everyday objects such as crayons and pencil-tip erasers. Place the different containers around the room to create estimation stations.

For the lesson, have partners travel to each station in your room, estimate the number of objects, and record their estimate.

After all estimates have been made, the class comes together. You show one of the collections and ask students about their estimates, recording different ideas on the board. Be sure to lift up estimates that are close together and those that are far apart. Do this for each collection.

STATION A	STATION B
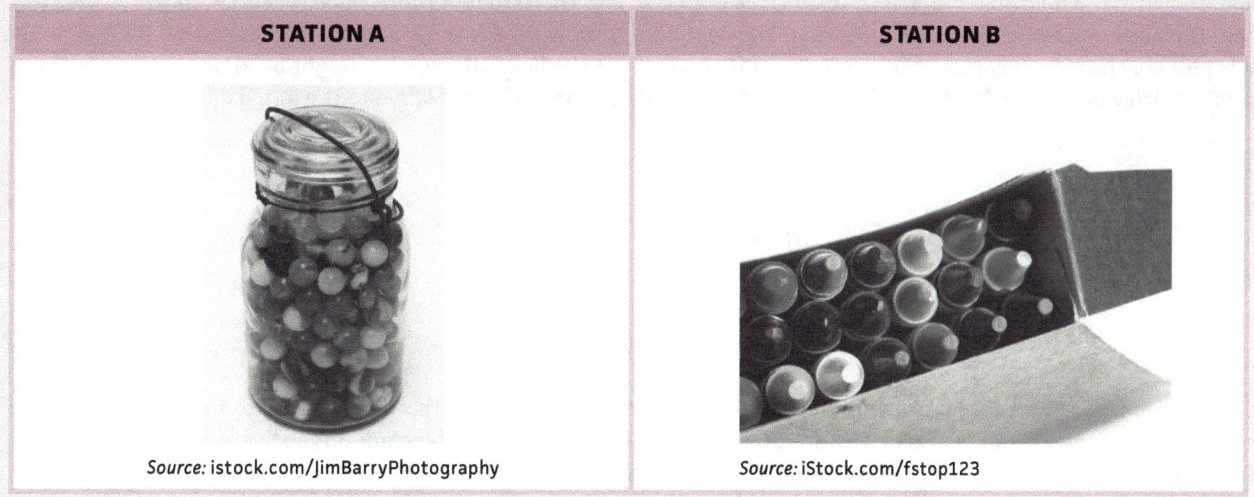	
Source: istock.com/JimBarryPhotography	*Source:* iStock.com/fstop123

After class estimates have been recorded, send students back to the stations. Have them count the exact number of objects (marbles, cubes, or crayons) that are in each collection. Note that you can choose to have each group count each collection or have a group count one collection and share their results. Bring the group together to compare the actual amount in each collection with student estimates. Help them know that 72 and 73 are both good estimates for a collection of 68 things. Stress that estimates are close and that guessing the correct number is all well and good but not the point of an estimate.

After providing this experience, you can work an estimation station into your daily class routines. It could be something done first or it could be part of a center or station. To minimize prep time, share estimation collections with your teammates and have a student create the estimation collection (count the objects and fill it) at the end of the day so it's ready for the next day.

ACTIVITY 1.2
NEAR AND FAR

Estimating helps students think about the *relative position* of numbers—how close or how far numbers are from one another. This allows students to think about numbers that are close to one another and those that aren't. By doing so, students will develop a better sense of what might be a good estimate: 50 is a better estimate for 45 than 80.

For this activity, gather a string and index cards. The string can be stretched out between two chairs by securing the ends or it can be attached to a wall or board. This string will serve as a number line where students will place number cards. Prepare number cards by folding each card in half and writing a number on each side (you may have heard this referred to as "clothesline math"). The numbers can be within any range that is appropriate for the students.

Place one card's fold over the string, so it hangs and shows the number. Start with a decade number—one ending in zero. Hold up another card and have students determine if it would be near or far from the first number. They will place it on the number line accordingly and provide their thinking about why they placed the card where they did. It might be easier for students to begin with numbers that are relatively close to each other. This allows students to visualize a number line. Shown in the following example, a 10 is placed on the string. Then students determine that 19 is far from 10, so this card is placed on the other end of the string. The next card is shown, a 12, and students determine that 12 is near 10. It is probably best to only place three or four cards at a time on the line. Otherwise, it might become too crowded with cards, which would make it difficult for students to think about the position relative to the initial number.

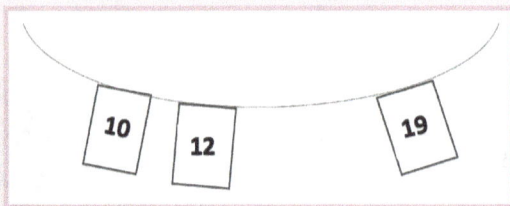

Continue to place different numbers on the number line and ask students, "Is this number near or far from the number that is on the number line?" Vary the position of where you hang the first number cards on the line, as this will encourage students to visualize a number line and think about the distance between numbers. It is important to note, however, that the exact spacing between the cards doesn't really matter for this activity. Some students may worry about the spacing and try to get it accurate, but that isn't the goal of this activity. Rather, it is to help them think about how close together or how far apart numbers are.

As a follow-up to a whole-class activity or station activity later, students can replicate this activity with partners. Students use a whiteboard to draw a line similar to the physical one used previously. One student places a number along the line and then gives the other student a number to position. The student who places the number says, "___ is near/far from ___." So, if a student writes 40 on the number line and offers the number 87, the second student places 87 further to the right of 40 and says, "87 is far from 40."

ACTIVITY 1.3
WHERE DOES IT GO?

Working with number lines is an excellent way for students to develop their sense of number relationships and estimation. However, experiences with number lines may all too often make use of "standard" endpoints. That is, students may typically work with number lines that have endpoints of 0 and 100, 0 and 1,000, 0 and 1 (with fractions), and so on. Because of this, students may develop incomplete or inaccurate ideas about how numbers are related (e.g., 50 is always in the middle and it's always half). This instructional activity is an entry point into a deeper understanding of number lines and the relative placement of numbers on them.

The sample set of number lines shown here are available online. They serve as a good starting point, but you should feel free to get creative. A good way to go about this work might be to have a variety of number line examples recreated on sentence strips and labeled with a letter. Then, give different examples to partners or small groups. Have students determine the endpoint for each number line before passing their number line on to the next group. Afterward, bring the whole class together to discuss their solutions and reasoning.

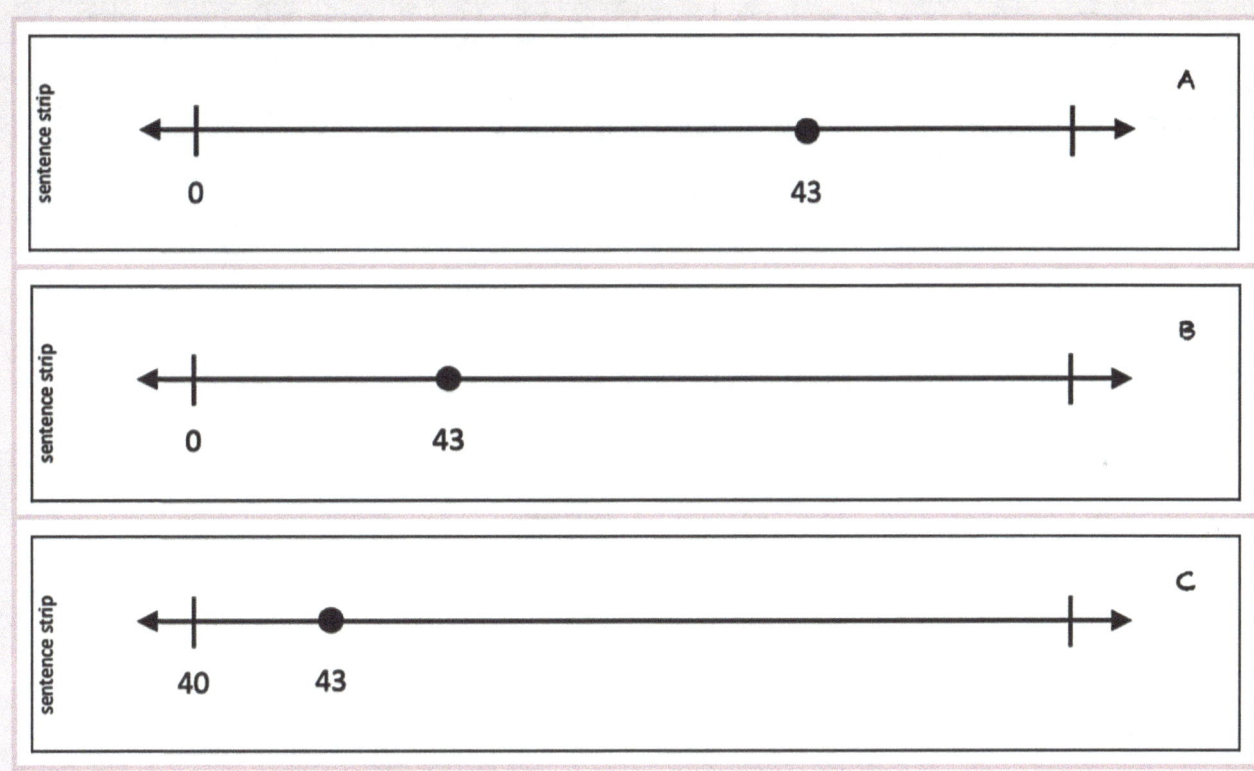

The sentence strip examples are one way to go about exploring numbers and number lines. Another way to explore this is by having students place a number on a number line (e.g., 82) and shift the location based on different knowns or different endpoints. The upper left shows a traditional placement example. The lower left shows endpoints of 50 and 100. On the right side, the location of 100 shifts while 0 remains the same.

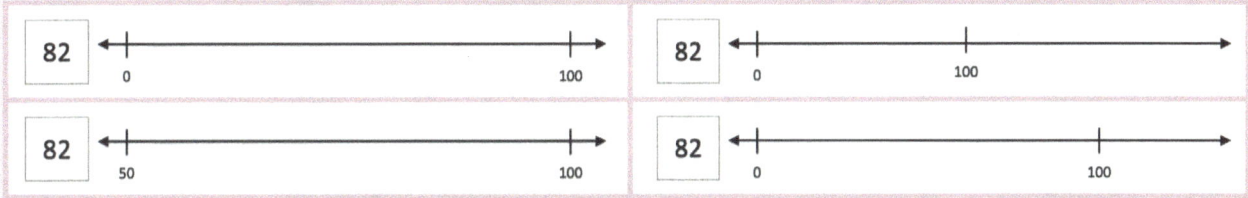

Students likely need many opportunities to explore, discuss, and practice the dynamic nature of numbers and number lines. An online resource, shown below, is available for you to edit and modify to create new independent practice exercises.

(Continued)

(Continued)

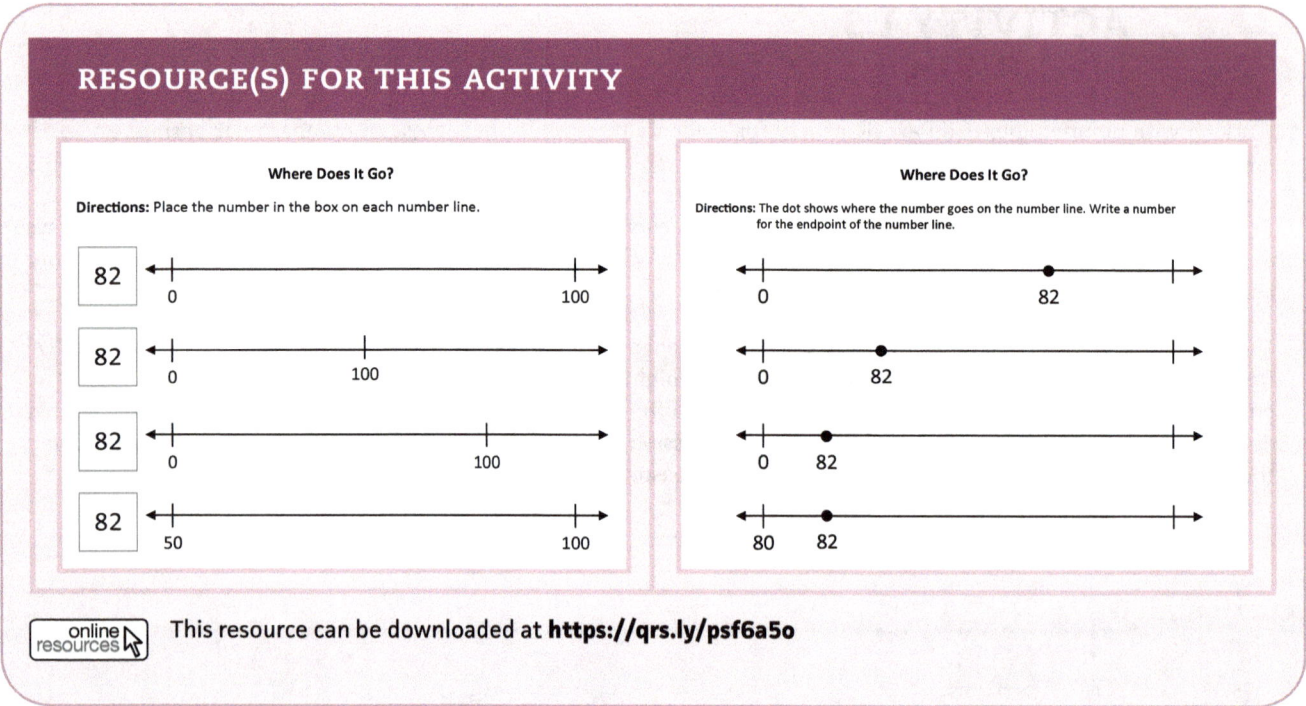

Where Does It Go?

Directions: Place the number in the box on each number line.

82

82

82

82

Where Does It Go?

Directions: The dot shows where the number goes on the number line. Write a number for the endpoint of the number line.

This resource can be downloaded at **https://qrs.ly/psf6a5o**

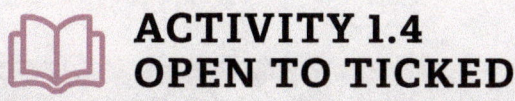

ACTIVITY 1.4
OPEN TO TICKED

Open number lines are invaluable for modeling computational strategies. But before students use them to show their strategies, they must understand them and how numbers are related on them. This is grounded in the understanding that number lines are dynamic, as described in Activity 1.3. This instructional activity complements that work, helping students recognize how open number lines work. You might use this activity before or after the previous lesson, based on the needs of your students.

In this activity, students see how an open number line connects with a ticked number line. For this activity, it is essential that students work with number lines that are the same length and that the partitions and number locations are similar. Begin by showing the open number line like the one at the top of the following image. Ask students where they would place different numbers, such as 7, 13, and 18, on that number line. Have them mark their locations and share their reasoning. Highlight how the numbers relate to 0 and 20 (i.e., 7 is closer to 0 or 18 is closer to 20). Then, ask them what number they think goes in the middle (in this case, 10 is in the middle).

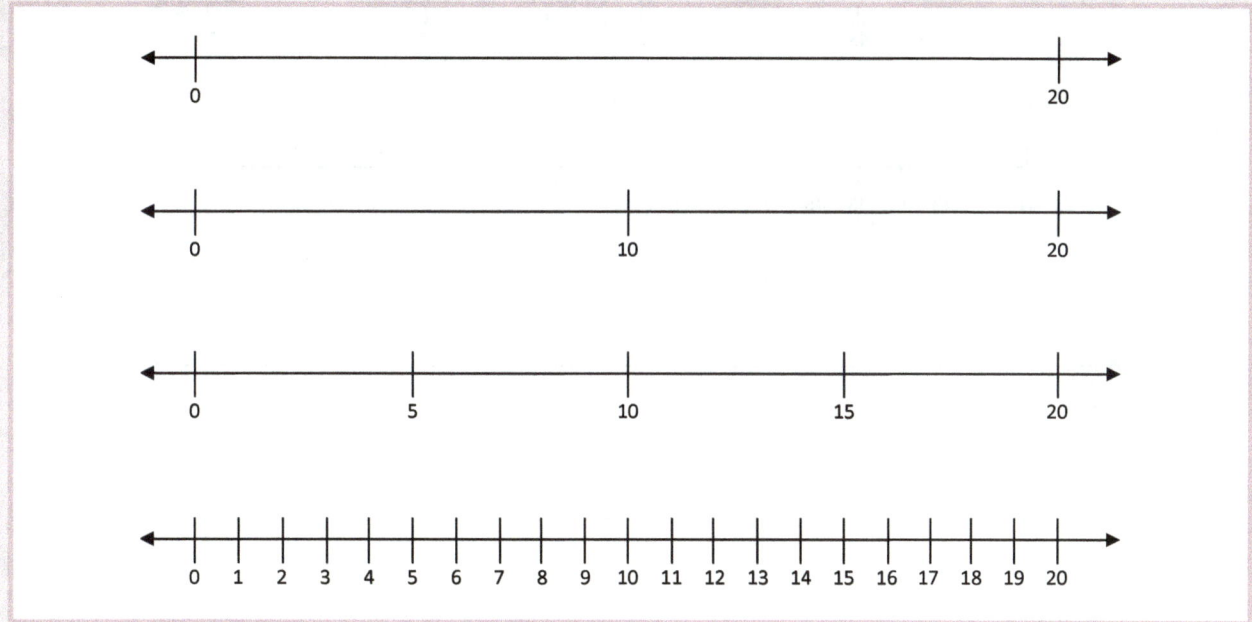

After discussion on the open number line, provide the next number line (second from the top) with 10 as the only tick mark. Ask them to place those same numbers on this new number line. Let them shift locations if they need to with this new information, again talking about relationships to given numbers. After work with the second number line, proceed to the third number line with the known locations of 5, 10, and 15. Finally, you want to introduce the last, fully ticked, number line. When this is introduced, compare and contrast the actual locations with their placements. Help them know that it's OK to be a little off. Repeat the experience by having them place different numbers.

You may be wondering if it's better to start with the ticked number line and move to the open one. While it may be helpful for some students, moving from open to ticked allows you to get a sense of how well they reason about number relationships and how well they estimate. That being said, it is perfectly reasonable to reverse the order of the number lines. What's most important is that students begin to understand how numbers are positioned on a number line.

The online resource provides a template for making number lines. The 0 to 20 lines, along with 40 to 60 and 0 to 100, are also available. As students show comfort and skill identifying placements of numbers between 0 and 20, begin to change the endpoints as in these examples; 0 to 100 number lines should be the last you introduce.

TEACHING TAKEAWAY

For helping students with decimal relationships, the 0 to 20 number line could be modified to show 0 to 2.0 and the 40 to 60 could be changed to 4.0 to 6.0.

(Continued)

(Continued)

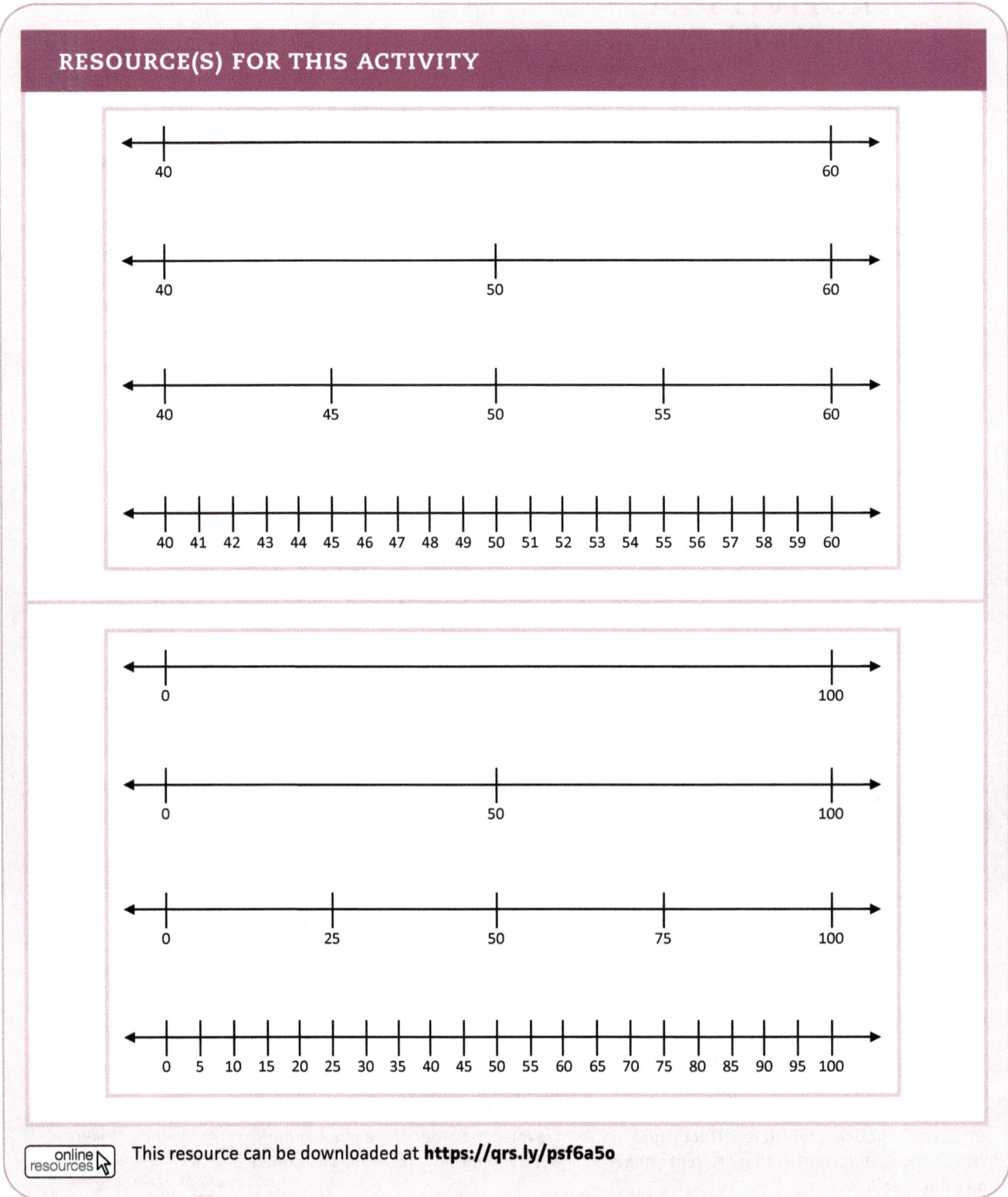

ACTIVITY 1.5
PROMPTS FOR RELATIVE SIZE OF A NUMBER

Use the prompts below as opportunities to develop understanding of and reasoning with the strategy. Have students use representations and tools to justify their thinking, including base ten models, number lines, number charts, and so on. After students work with the prompt(s), bring the class together to exchange ideas. Remember that these could be useful for collecting evidence of student understanding.

- Eddie says that 27 is close to 100 because it has a 7 in it. State why you agree or disagree with Eddie.

- Create a number close to 0, a number close to 50, and a number close to 100 that each have a 6 in them. Use pictures and numbers to prove your thinking.

- Ebony says that a number moves on a number line when the endpoints of the number line change. Do you agree or disagree with Ebony? Create an example to show your thinking.

- Jake says that 49 is close to four different numbers. What numbers do you think those are? Explain your thinking.

- What are some numbers that 83 is close to? What are some numbers that 83 is far from?

- Krissy thinks that two-digit numbers greater than 20 are all close to 100. Why do you agree or disagree with Krissy?

- Sometimes a number is big and sometimes it is small. When is 50 a big number? When is 50 a small number?

- State how you know if a number is close to or far from another number. Create some examples to show your thinking.

QUALITY PRACTICE FOR NUMBER RELATIONSHIPS

You have to vary how students practice to maximize the quality and effectiveness of that practice (Bay-Williams & SanGiovanni, 2021). The practice types are varied. You'll find routines that are perfect for student reasoning and discourse. They promote the exchange of ideas and help students better understand the featured fluency foundation. Games play out differently. They are opportunities to sharpen one's automaticity with a foundation through joy and play. Centers are independent practice endeavors that can be reused. They mix in processing and reflection. But know that the different activities vary the aspects of each module's foundation so that students are skilled with all of them.

ACTIVITY 1.6

Name: "Dynamic Number Line"   **Type:** Routine

About the Routine: A number's position on a number line is dynamic. It changes as the endpoints on the number line change. This routine is an opportunity to deepen students' understanding of the relationship between numbers through number lines. Early use with this routine may call for using two number lines instead of three. This routine, like many others in this book, is a perfect follow-up experience for practicing concepts and skills developed in the instructional activities (e.g., Activity 1.3 and Activity 1.4).

Materials: There are no materials needed for this routine.

Directions: 1. Post a number line with given endpoints and a point on the number line.

2. Student partners discuss what value they think the point represents.

3. The class discusses their ideas and comes to an agreement.

4. Then, draw a second number line directly below the first number line with a point placed at the same location as the original number line. This second number line has new endpoints, but the point is in the same location on the line.

5. Students first determine the value of the point on this second number line and then identify where the value from the first number line belongs on this second line.

6. Draw a third number line with yet another change in the endpoints.

7. Students again determine the value of the point on this third number line and then identify where the values from the first two number lines belong on this third line.

In this example, the teacher started with a number line having endpoints of 0 and 50. Students discussed. Some thought it was 40, some thought 41, others thought 42. The group settled on 42.

Then, the next number line was presented. The teacher showed where 42 is when placed on this new number line. Now, students discussed what the point now shows. After discussion, the class agreed to think of it as 74.

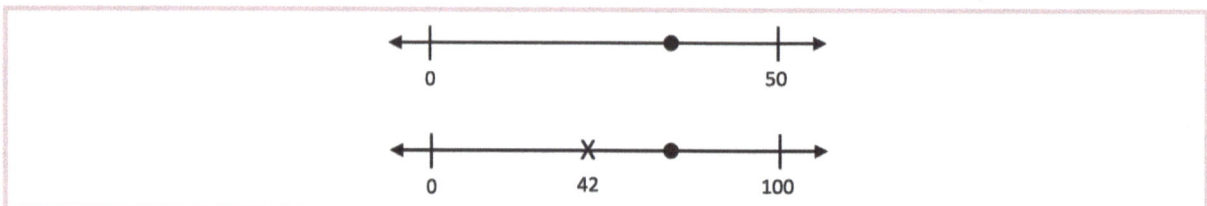

The teacher provided the last number line, now with endpoints of 0 and 200. They recorded where 42 and 74 were placed. Students talked about what the point might be now.

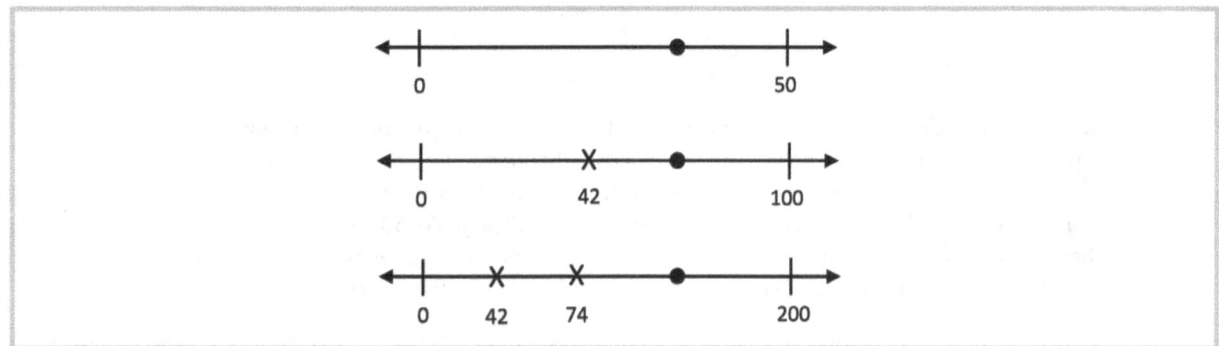

An alternate version of this routine is laid out in the following images. With it, the endpoint remains constant. In this version, students reason about a given point. Discussion is had and an agreement is made. During that conversation, acknowledge that a few different numbers are reasonable, but students need to agree on one choice.

Assuming the class agreed with 31 for the first number line, a new point is identified and the discussion continues.

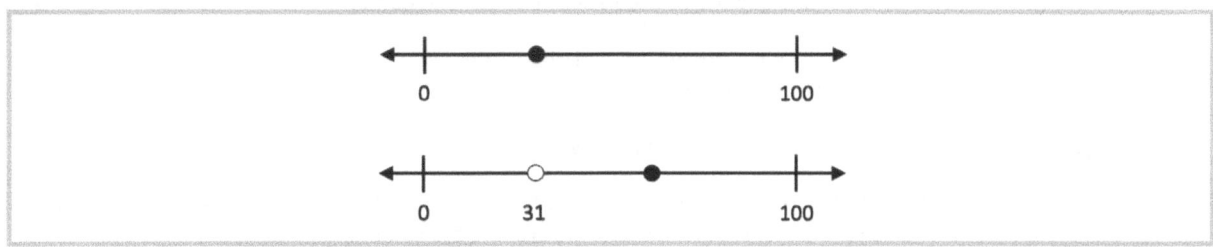

Finally, a third point is presented and discussion is had. While this is shown on three different number lines, it is perfectly fine to present all three points on the same number line.

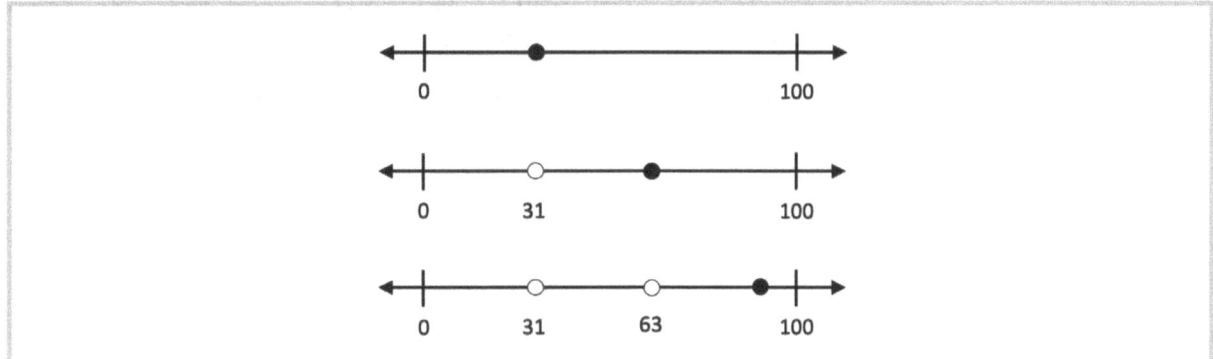

ACTIVITY 1.7

Name: "The Stand"   **Type:** Routine

About the Routine: The Stand is a playful routine perfect for practicing concepts such as number relationships. It plays out like the iconic television show *Survivor*. Students who meet conditions "survive" the round, but those who don't may be better positioned to win in the long run. As with other "close to" games, begin with benchmarks (e.g., 0, 50, and 100). In time, transition to multiples of 10 (e.g., 20, 30, or 70), multiples of 5 (e.g., 35, 65, or 85), and then any two-digit number (e.g., 37, 53, or 71). In later grades, students can make three-digit numbers by dealing themselves four cards and even four-digit numbers by dealing five cards.

Materials: Two sets of digit cards (0–9) per student or a deck of playing cards (aces = 1, remove tens and face cards)

Directions:

1. Each student stands at their desk.

2. Call a "close to" situation. For example, you might say, "Make a number close to 50."

3. Students then use two of their three cards to make a two-digit number close to the target you call.

4. Call two students at random.

   a. The student closest to the target remains standing. The other student sits down, but they are not out of the game.

   b. If both students are within 5 of the target, they remain standing regardless of who is closer.

   c. Repeat by calling on two more students and then again with two more students.

5. After calling on three sets of two students, have all students clear their cards and deal themselves three new cards. Students who are standing stay standing.

6. Call a new "close to" situation (e.g., close to 25).

7. Call on two students at a time again with one notable change: If you call on a student who is seated and they are closest to the target or within 5 of it, then they get to stand up and pick two classmates to sit down regardless of how close to the target those classmates are.

8. Continue playing. The goal of the game is to be the survivor—the last person standing.

> **TEACHING TAKEAWAY**
>
> Competitive routines such as this can increase engagement for some but not others, who may get anxious. In this case, pair students to collaborate on their number-making.

ACTIVITY 1.8

Name: Number Line Cross Off   **Type:** Game

About the Game: Number lines help students see relationships between numbers. They show how close numbers are to benchmark numbers as well as how far two numbers are from one another. As you know, students need lots of exposure to number lines before using them becomes second nature. This game is an opportunity for that sort of practice. In it, students take turns creating numbers and spotting them on a number line. It's important to note that students shouldn't be allowed to reverse digits until you are confident in their understanding of place value. For example, a student who rolls an 8 and a 2 should only use it as 82. In time, they can use it as 82 or 28, but you don't want to permit this during early place-value instruction and practice.

Materials: Playing cards (queens = 0, aces = 1; remove tens, kings, and jacks) or a 10-sided die; one *Number Line Cross Off* game board per two players

Directions: 1. Each player chooses eight sections on the number line. A section is a part of the number line between a large tick mark (e.g., 0, 10, 20, etc.) and a small tick mark (e.g., 5, 10, 15, etc.).

2. Players then take turns generating two-digit numbers with the playing cards or die. If a player's number falls within an identified section of their opponent's number line, they cross off the section. If it doesn't fall within a section or if that section is already crossed off, they lose their turn.

3. The first player to cross off all eight of their opponent's sections wins the game.

In the example, Player 1 chose the sections 10–15, 30–35, 45–50, 60–65, 65–70, 75–80, 90–95, and 95–100, as shown by the rectangles drawn above each section. Player 2's sections are recorded on the second number line. Player 1 rolled 41 on their first turn and crossed off the 40–45 section on Player 2's number line. Player 2 then rolled a 32 and crossed off Player 1's section between 30 and 35. On Player 1's next turn, they rolled a 94 but they lost their turn because Player 2 didn't identify the section corresponding to 94 on their number line.

> **TEACHING TAKEAWAY**
>
> Most games in this book come with ready-to-print game boards for use in your classroom, but many of them can be made easily on a piece of notebook paper. Show your students how to do this so that they can play the same game at home with family.

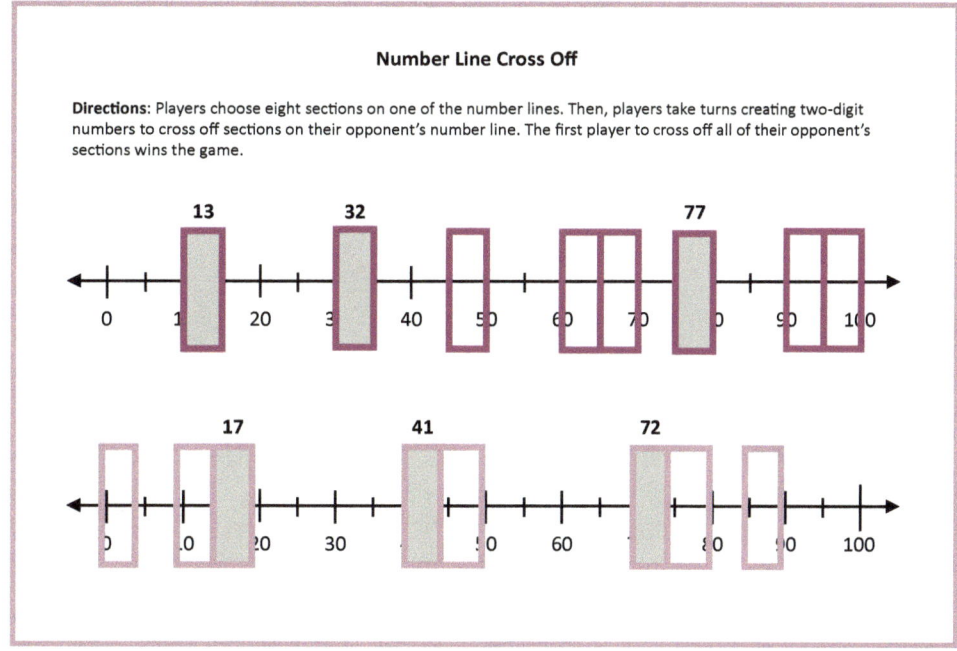

Number Line Cross Off

Directions: Players choose eight sections on one of the number lines. Then, players take turns creating two-digit numbers to cross off sections on their opponent's number line. The first player to cross off all of their opponent's sections wins the game.

(Continued)

(*Continued*)

This version of *Number Line Cross Off* features endpoints of 0 and 100. It can be easily adapted for endpoints of 0 and 1,000 with tick marks representing 50s and 100s (left image). To play, students would need to generate three-digit numbers. You can also modify it to have endpoints of 0 and 1.00 or 0 and 10.0 with tick marks representing 0.5 and 1.0. You can create number lines with unique endpoints (e.g., 600 and 700). Doing the latter can help students see connections among all sorts of numbers.

RESOURCE(S) FOR THIS ACTIVITY

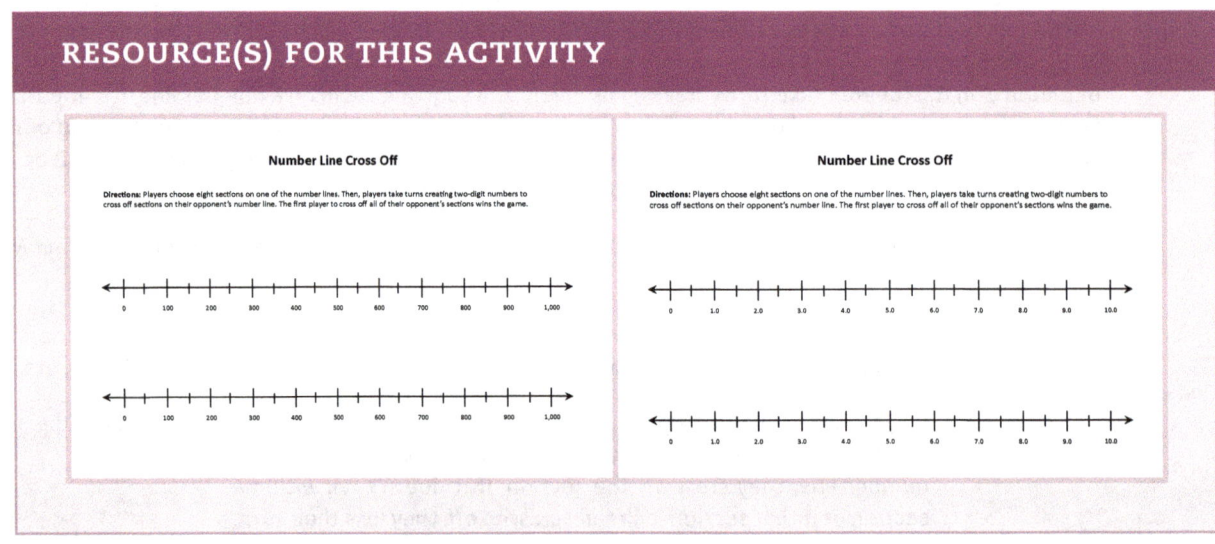

online resources → This resource can be downloaded at **https://qrs.ly/psf6a5o**

ACTIVITY 1.9

Name: Five Targets   **Type:** Game

About the Game: Target games are perfect for practicing number relationships. Students can play games with the goal of getting as close to a target as possible, such as in this game, *Five Targets*. You can easily change the goal of these games to have students get as far away from a given target as well. You can provide number charts and calculators to help students determine exactly how far from a target their numbers are. Over time, you can change questions from asking "how far from a target" to "about how far," allowing students to practice estimation. For example, a student who makes 33 for a target of 100 can say their number is about 70 from 100. You might be thinking that this game will be rough because your students are challenged to add all of the differences. Have them use a calculator! Or have them predict who won by examining and comparing the differences. They might not even need to add all of the differences. In the example, Player 1's difference of 144 is more than all of Player 2's differences added together.

Materials: Playing cards (queens = 0, aces = 1; remove tens, kings, and jacks) or a 10-sided die; one copy of *Five Targets* game board per two players

1. Using the die or cards, players take turns generating digits.

2. Players place digits in the spaces on their side of the game board with a goal of making a completed number that is as close to the target as possible.

3. Players are able to discard digits two times.

4. After all spaces are completed, players find how close they are to each target.

5. The player who is closest to the most targets wins the game.

The example shows a completed game board. You can see that Player 2 on the right was closer to four of the targets (10, 50, 100, and 100) and won the game. You might be wondering about why certain digits were or weren't discarded. Remember, players have to place a digit in one of the spaces before the next turn and once placed, a digit can't be moved. So, either of Player 2's discards would have been better choices than the 8 in the last target, but those numbers came up after the player had already placed the 8.

> **TEACHING TAKEAWAY**
>
> As you will see in the next activity, you can change targets from common benchmarks such as 100 to any number, such as 87 or 93. This provides a good challenge for honing student skills with number relationships.

Five Targets

Directions: Players take turns placing digits in open spaces to create numbers as close to the target as possible. A player may discard up to two digits. After all spaces are filled, players determine how close they are to each target. The player who is closest to a target the most times wins.

Player 1		Target	How close?		Player 2		Target	How close?
1	5	10	5		0	6	10	4
3	1	25	6		1	9	25	6
6	3	50	13		5	3	50	3
8	7	100	13		9	8	100	2
2	4	4	100	144	1	1	5	100	15
Discards →	8	8			Discards →	3	2		

(Continued)

(Continued)

You could change the targets to tens (20, 30, 40, 50, 60), fives (25, 35, 45, 55, 65), or random two-digit numbers (37, 53, 62, 71, 94). The game also works well with three-digit numbers and decimals in later grades, as shown.

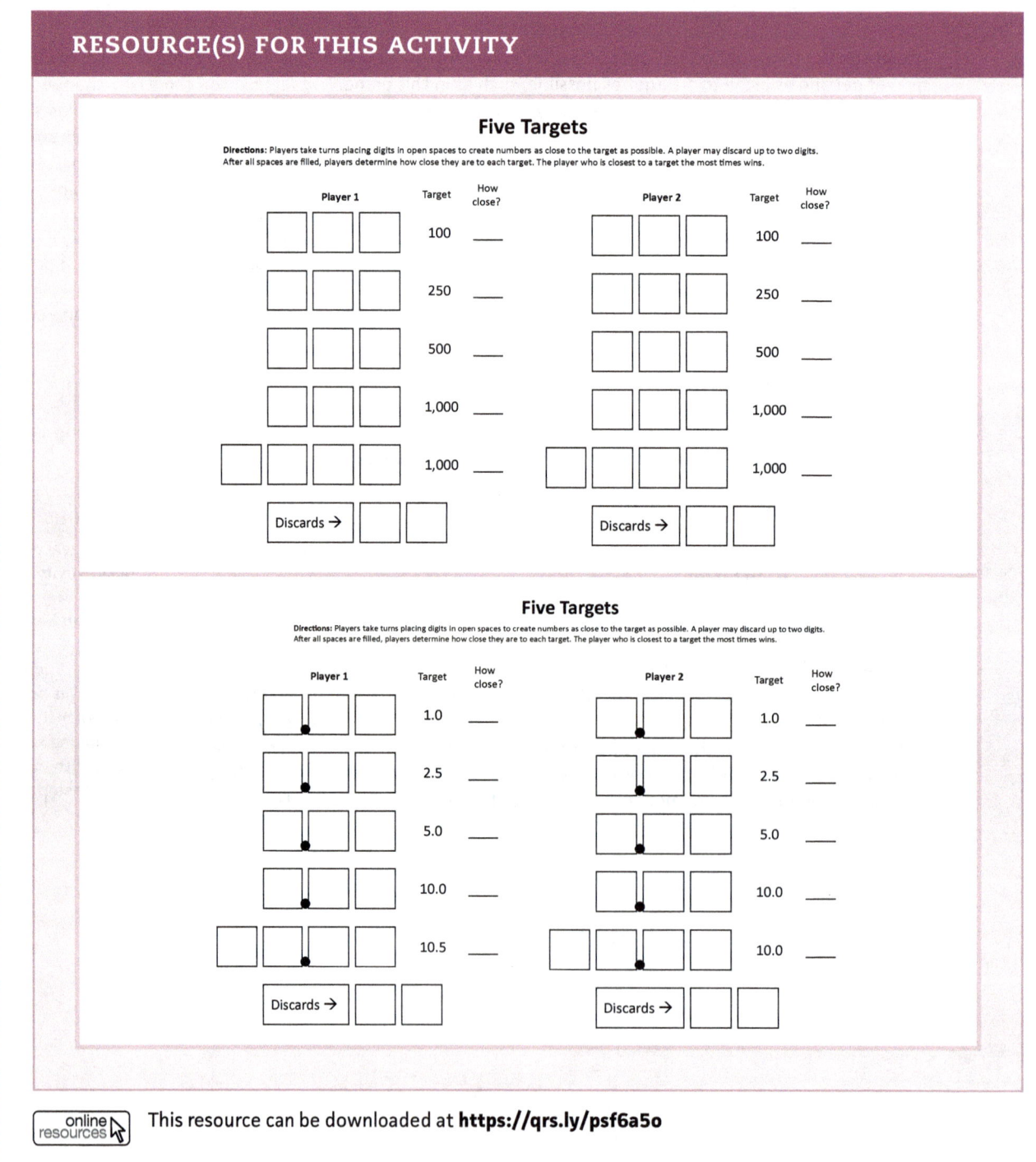

RESOURCE(S) FOR THIS ACTIVITY

Five Targets

Directions: Players take turns placing digits in open spaces to create numbers as close to the target as possible. A player may discard up to two digits. After all spaces are filled, players determine how close they are to each target. The player who is closest to a target the most times wins.

online resources — This resource can be downloaded at **https://qrs.ly/psf6a5o**

ACTIVITY 1.10

Name: Close To   **Type:** Game

About the Game: Relating numbers to benchmarks is useful for comparison and reasoning about their distance from other numbers. This reasoning is helpful for making decisions about which number to decompose when computing or which number to start with when adding, and there are obvious applications to determining the reasonableness of solutions. In this game, players take turns making numbers and determining who is closer to a given benchmark.

Materials: Playing cards (queens = 0, aces = 1; remove tens, kings, and jacks) or a 10-sided die; deck of *Close To* game cards

Directions: 1. The *Close To* game cards are shuffled and placed face down.

2. A game card is flipped over.

3. Both players make a two-digit number with their playing cards or die.

4. The player closest to the number on the game card takes the card. If both players' numbers are equally close to the number (e.g., 32 and 68 are both 18 from 50), the next card is flipped and the player closest to that card takes both cards.

5. The player with the most cards at the end of the game wins.

Close To game cards are available as an online resource. The images show those cards along with variations of the game. The second set of cards shows a version of the game in which students compare their numbers to numbers that aren't benchmarks. The third image shows how the game can be played with decimals. You could easily modify it for three- and four-digit numbers as well. The right image shows a set of number lines. This tool can support students' reasoning as their understanding and skill with relationships develops. Simply have them record 0, 50, and 100 on each given tick mark. Then, they can place an *x* where they think their number would go and another letter or symbol where they think their opponent's number would go.

TEACHING TAKEAWAY

Print two copies of recording sheets (the blank number lines) and create two-sided copies. Then laminate the two-sided copy so that the resource can be used many times.

RESOURCE(S) FOR THIS ACTIVITY

Closer to 0	Closer to 50	Closer to 100
Closer to 0	Closer to 50	Closer to 100
Closer to 0	Closer to 50	Closer to 100
Closer to 0	Closer to 50	Closer to 100
Closer to 0	Closer to 50	Closer to 100

Closer to 25	Closer to 75	Closer to 85
Closer to 90	Closer to 35	Closer to 43
Closer to 16	Closer to 88	Closer to 39
Closer to 84	Closer to 77	Closer to 29
Closer to 51	Closer to 63	Closer to 47

Closer to 0	Closer to 0.5	Closer to 1.0
Closer to 0	Closer to 0.5	Closer to 1.0
Closer to 0	Closer to 0.5	Closer to 1.0
Closer to 0	Closer to 0.5	Closer to 1.0
Closer to 0	Closer to 0.5	Closer to 1.0

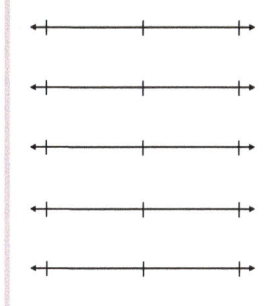

online resources ⇗ This resource can be downloaded at **https://qrs.ly/psf6a5o**

ACTIVITY 1.11

Name: Ten Close Calls   **Type:** Game

About the Game: *Ten Close Calls* is another game for determining whether numbers are close to others. Before playing this or any other "close to" game, you want to have discussions with students about examples of numbers that are close to one another. For two-digit numbers, close numbers are usually within 2 or 3 of the number (e.g., 67 is close to 70). You can extend the limit to within 5 if appropriate. For three-digit numbers, close numbers should be within twenty or so of the given number (e.g., 678 is close to 700). Of course, these are not rules. You can start with a range and narrow it as students become more skilled.

Materials: Playing cards (queens = 0, aces = 1; remove tens; kings and jacks are wild) or a 10-sided die; one *Ten Close Calls* game board per two players

Directions: 1. Players take turns making a three-digit number by dealing themselves three cards. Note that dealing a 0 as the first digit creates a two-digit number (e.g., 0, 3, 5 makes 35). Optional: Players can rearrange their cards after dealing them (e.g., 0, 3, 5 can be 35, 530, or 350).

2. Players place the number they make in a space closest to the given number. However, the number must be within 25 of the hundred. Note: This is a suggestion; "within 25" can be adjusted as needed.

3. If a player cannot place their number, they lose their turn. Players might not be able to place their number because it isn't within 25 of any hundred or because a space is already filled.

4. The first player to fill all of their spaces wins the game.

You can modify *Ten Close Calls* in the same way(s) that you might modify the previous game. Students can make two-digit numbers trying to get within 2 or 3 of a 10 (first example). They can make three-digit numbers as described in the directions. They can try to get within a few of a 100 and 50 (third example) or completely random numbers. Of course, these are only some beginning ideas. This game can be adapted to decimals and fractions. Another alternative is to have ten empty targets and have students make random numbers with cards before trying to make close to numbers.

> **TEACHING TAKEAWAY**
>
> Any game can be turned into a center by having students try to complete the task in the fewest number of tries or finding the smallest total difference.

In this example, players took turns rolling two-digit numbers and placing them on the board. In this version, players were not allowed to reverse digits. On Player 1's first turn, they rolled 49 and decided it was close to 50. Player 2 has only two numbers on their board because on their third turn they rolled 41 and the space for close to 40 was already filled. Also note that they used 15 as close to 10, which is fine; it could also have been used as close to 20. We suggest that as students become more experienced, their numbers should be within 2 or 3 of the target.

Ten Close Calls

Directions: Players take turns making numbers. They put their number in the space that their number is closest to. If a player makes a number that isn't close to any number or the space is filled they lose their turn. The first player to fill all of their spaces wins the game.

Player 1		Player 2	
10		10	15
20		20	
30	32	30	
40		40	43
50	49	50	
60	63	60	
70		70	
80		80	
90		90	
100		100	

Ten Close Calls

Directions: Players take turns making numbers. They put their number in the space that their number is closest to. If a player makes a number that isn't close to any number or if the space is filled, then they lose their turn. The first player to fill all of their spaces wins the game.

Player 1		Player 2	
50		50	
100		100	
150		150	
200		200	
250		250	
300		300	
350		350	
400		400	
450		450	
500		500	

Ten Close Calls

Directions: Players take turns making numbers. They put their number in the space that their number is closest to. If a player makes a number that isn't close to any number or if the space is filled, then they lose their turn. The first player to fill all of their spaces wins the game.

Player 1		Player 2	
24		24	
37		37	
43		43	
48		48	
52		52	
64		64	
76		76	
81		81	
88		88	
93		93	

Ten Close Calls

Directions: Players take turns making numbers. They put their number in the space that their number is closest to. If a player makes a number that isn't close to any number or if the space is filled, then they lose their turn. The first player to fill all of their spaces wins the game.

Player 1		Player 2	
10		10	
20		20	
30		30	
40		40	
50		50	
60		60	
70		70	
80		80	
90		90	
100		100	

Ten Close Calls

Directions: Players take turns making numbers. They put their number in the space that their number is closest to. If a player makes a number that isn't close to any number or if the space is filled, then they lose their turn. The first player to fill all of their spaces wins the game.

Player 1		Player 2	
100		100	
200		200	
300		300	
400		400	
500		500	
600		600	
700		700	
800		800	
900		900	
1,000		1,000	

online resources — This resource can be downloaded at **https://qrs.ly/psf6a5o**

ACTIVITY 1.12

Name: The Sort　　　　　　　　　　　　　　　　**Type:** Center

About the Center: The Sort is a simple center where students generate numbers and sort them into groups based on their proximity to a benchmark. But it does offer a slight twist for the individual student: In short, a student sets a goal for filling one of the columns ("close to" numbers) within a given number of turns. The twist adds an element of chance and fun. The center can be adjusted for a variety of benchmark numbers such as those in the example. You can adjust the center to practice three-digit numbers (benchmarks of 0, 250, 500, 750, etc.) or, in later grades, students can practice with fractions. Remember that you will need to establish what qualifies as "close." At first, it might simply be the closest benchmark (e.g., 41 is closer to 50 than 0 or 100). This version is shown in the example. In time, you might say that a number has to be within 5 or so to be considered close. Also note that resources like open number lines can be helpful tools for supporting student reasoning.

Materials: Playing cards (queens = 0, aces = 1; remove tens; kings and jacks are wild) or a 10-sided die; a copy of The Sort recording sheet (optional)

Directions:　1. The student sets a target for filling a "close to" column by completing the statement, "I can fill the close to ___ column in ____ turns."

　　　　　　　　2. Using playing cards or the die, the student generates a number and determines which column it belongs in.

　　　　　　　　3. The student adds a tally mark to the "number of turns box" on the recording sheet.

　　　　　　　　4. The student repeats the process until the target column is filled.

　　　　　　　　5. The student determines how close they were to the number of turns it took.

In the example, students were assigned "close to" numbers of 0, 50, and 100; these numbers can be changed as needed. This student selected the Close to 50 column and said that they could fill the column in ten turns. You can see the numbers they created and that it took twelve tries or turns to fill the column, as shown by the tally marks. Note that the student made 6 by rolling a 0 and a 6. You can choose to require students to use the digits as they appear (e.g., 1 and 7 makes 17) or allow them to rearrange the digits (e.g., 1 and 7 can be 17 or 71).

The Sort

I can fill the Close to 50 column in 10 turns.

Close to 0	Close to 50	Close to 100
6	44	98
13	53	93
22	55	85
8	61	
	69	

Number of Turns:

꜔꜔꜔꜔꜔　꜔꜔꜔꜔꜔

| |

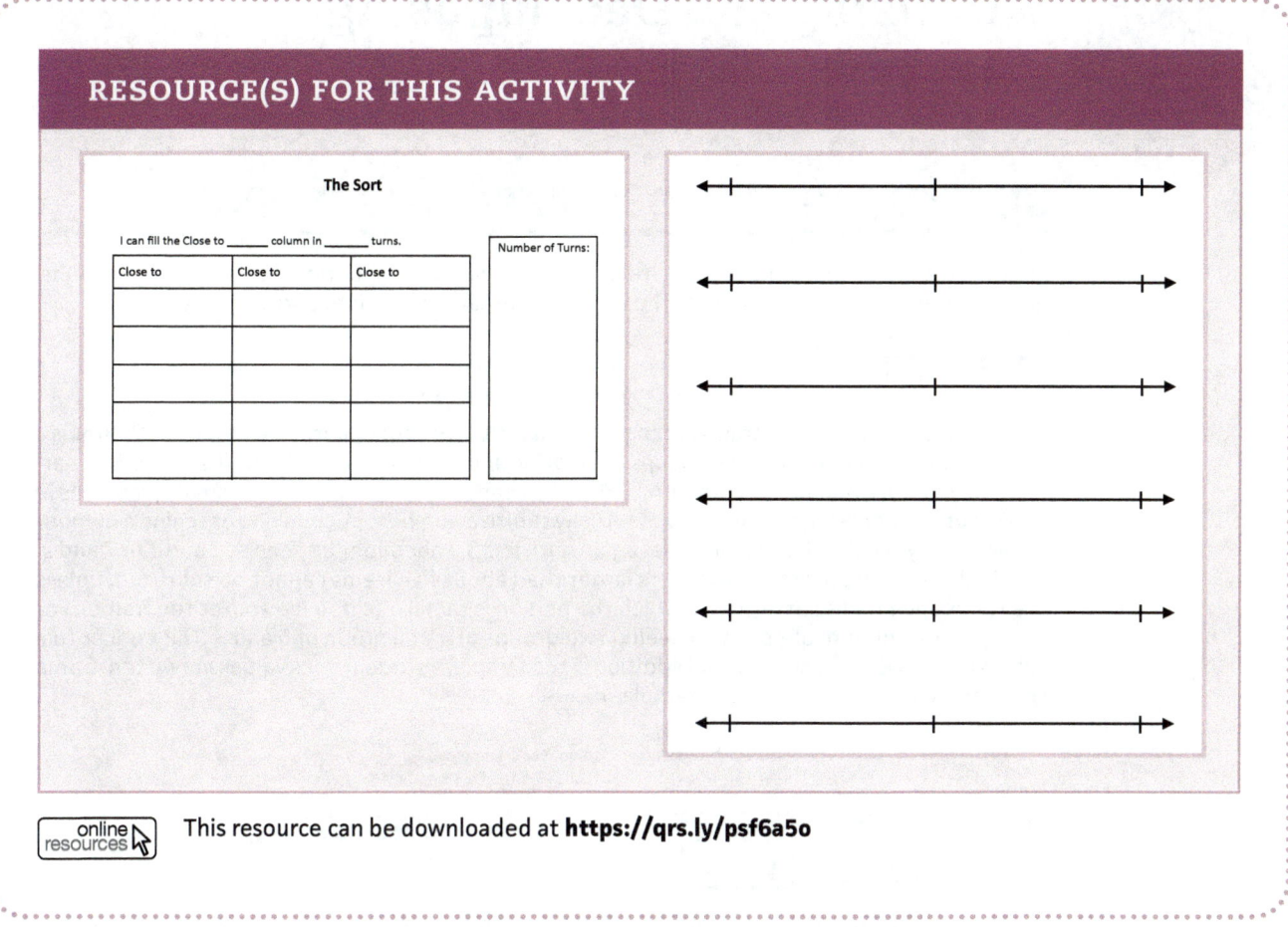

The Sort

I can fill the Close to _____ column in _____ turns.

Close to	Close to	Close to

Number of Turns:

online resources

This resource can be downloaded at **https://qrs.ly/psf6a5o**

NOTES

_____

_____

_____

_____

_____

_____

_____

_____

Subitizing and Decomposing

SUBITIZING AND DECOMPOSING OVERVIEW

Subitizing and decomposing are interrelated ideas that are foundational to all four operations, helping students to see the parts of the whole and break a whole into parts flexibly.

SUBITIZING

Subitizing is the ability to instantly recognize a quantity without counting (Clements, 1999). It is a powerful number concept that develops estimation and computation skills. At first, students perceptually subitize numbers such as 3, 4, and 5. They can simply recognize these small quantities in one group. Students then begin to conceptually subitize numbers by chunking or seeing a number in parts. They see 2 and 1 over and over again until it is 3. Four might be seen as 3 and 1 or 2 and 2. Five is also seen in different ways. Sets larger than about five items cannot be subitized, unless the items appear in a pattern with which the person is familiar (e.g., a die and/or ten frame). For example, a student might see 8 by seeing two groups of 4 or a group of 5 and 3. This conceptual subitizing grows into counting and addition. For example, a student sees values in two ten-frame spots and sees the quantity (for example, 10 + 3).

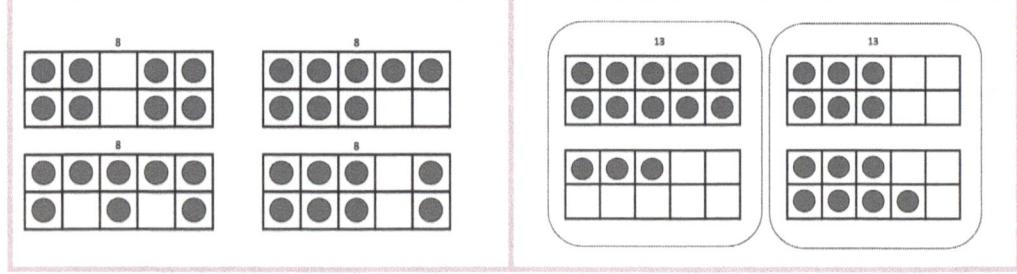

DECOMPOSING

Decomposing is the ability to separate or break apart a whole (of any size) into two or more parts. For example, for the number 13, you can decompose it into 10 and 3 (decompose by place value) or other ways (e.g., 6 and 7, or 5 and 6 and 2). In early grades, students learn to decompose numbers to ten in different ways, starting with physical representations. In later grades, students decompose multi-digit numbers in different ways. While decomposing by place value is common, students need significant experiences decomposing in other ways to support their reasoning. For example, the Make Tens strategy utilizes decomposing and what gets decomposed depends on the numbers in the problem. For 689 + 745, for example, you can decompose 745 into 734 + 11 and add the 11 onto 689 to create the equivalent expression 700 + 734 (Make Hundreds). This applies to fractions and decimals as well. To add 6.7 + 8.8, you can decompose 6.7 into 6.5 + 0.2; add the 0.2 to the 8.8 and you have the equivalent expression 6.5 + 9 (Make a Whole). Decomposing activities include using part–part–whole placemats and number bonds as students break apart a whole into parts.

Subitizing and decomposing are necessary foundations for counting and adding. While decomposing is commonly named in kindergarten standards, the skill is necessary across the grades to support computational fluency. This module provides a wealth of activities that can be implemented across grades either for intervention or as a foundation-building activity prior to introducing the Make Tens (or Make a Whole) strategy. Students might show capability with both, but their skills might be fragile. In other instances, neither idea has fully taken root before the curriculum has moved on. This module provides ample resources for revisiting these.

HOW DO SUBITIZING AND DECOMPOSING CONTRIBUTE TO FLUENCY STRATEGIES?

Decomposing numbers almost always plays a role in computational fluency. Consider 36 + 28: To use partial sums, you have to be able to decompose both addends by place value as shown in this first example.

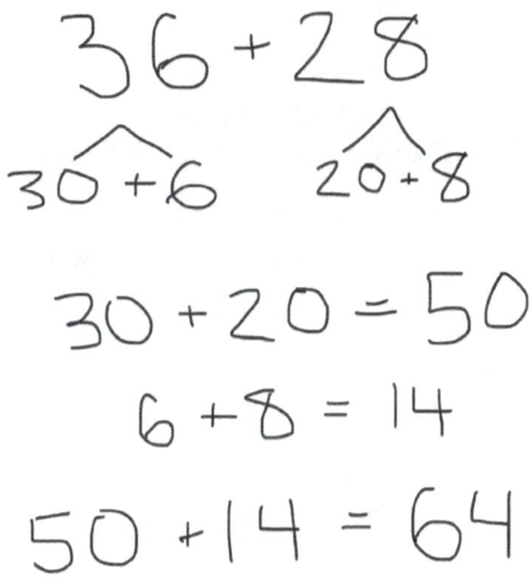

$$36 + 28$$
$$30 + 6 \qquad 20 + 8$$

$$30 + 20 = 50$$
$$6 + 8 = 14$$
$$50 + 14 = 64$$

To use a Make Tens strategy, you decompose one addend so that you can make a ten with the other number. In this problem, 36 and 28 can be decomposed. In the left example, 36 is decomposed into 34 and 2 to join the 2 with 28 to make 30. In the right example, 28 is broken into 4 and 24 so that 4 can be joined with 36 to make 40.

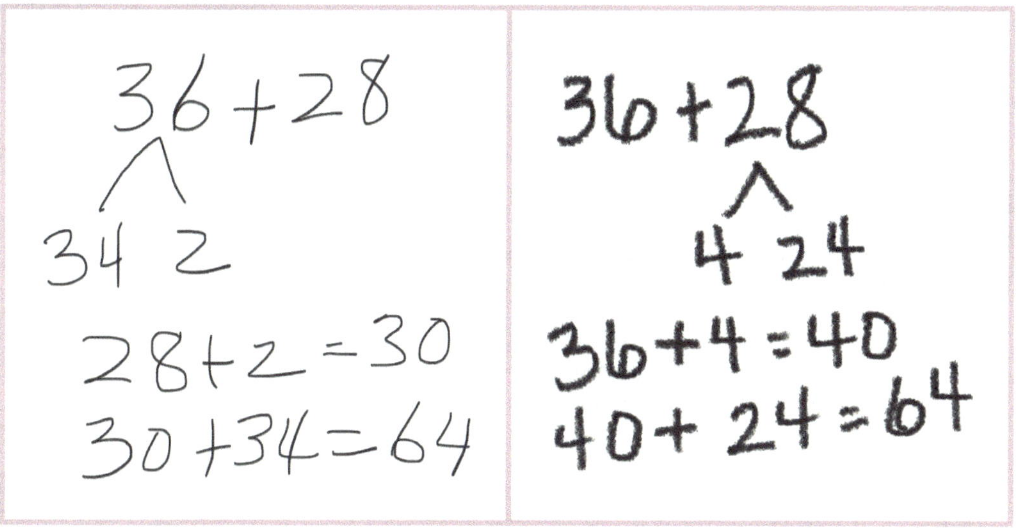

$$36 + 28$$
$$34 \quad 2$$
$$28 + 2 = 30$$
$$30 + 34 = 64$$

$$36 + 28$$
$$4 \quad 24$$
$$36 + 4 = 40$$
$$40 + 24 = 64$$

Counting On calls for flexible decomposition as well. Instead of breaking a number apart using place value, a student will need to decompose it into chunks for convenient counting. This will be rooted in the student's ability to skip count (Module 4). In this example, the student isn't able to count on by 20; instead, they count on by tens and then ones.

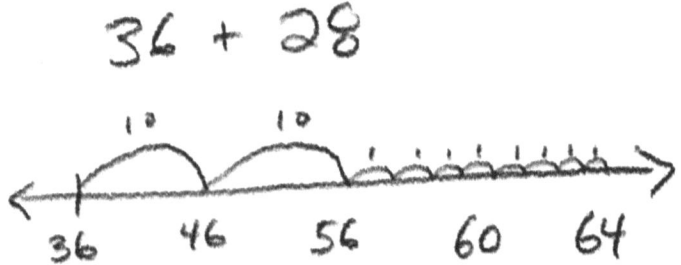

The subitizing and decomposing reference page provides an overview, important representations, connections to fluency, and actions to avoid. It can be downloaded and used for reference in planning, teaching, and discussions with colleagues and families.

FOUNDATION 2: SUBITIZING AND DECOMPOSING

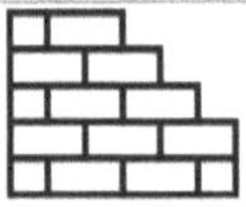

Overview	Subitizing is the ability to recognize a quantity without counting. Decomposing is the ability to separate or break apart a whole (of any size) into two or more parts.
Important Representations and Tools	**Dot Patterns** **Example:** Giving quick looks of a dot pattern on a paper plate or ten-frame, then asking, "how many dots do you see?" **Rekenrek** **Example:** Moving beads to the left and asking, "how many?" **Linking Cubes and Base 10 Pieces** **Example:** Using a Part-Part-Whole Placemat, decide what parts will equal the whole. **Number Lines** **Example:** Showing different jumps to get to 16. **Number Bonds** **Example:** Showing different ways to decompose 47.
Connection to Fluency	Subitizing leads to decomposing, and decomposing is necessary for many reasoning strategies, for example: **Make Tens:** to add 37 + 55, decompose 55 into 3 + 52 to get 40 + 52 **Break Apart to Multiply:** to multiply 12 × 9, decompose 12 into 10 + 2, and multiply each part by 9: 90 + 18 = 108
Actions to Avoid	• Under-utilizing representations – notice how many options you have! • Limiting decomposing to just one way (i.e., by place value)

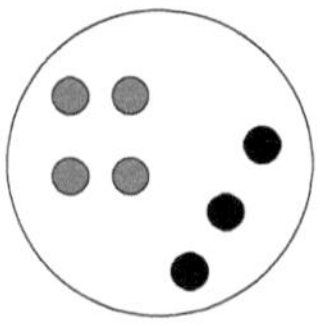

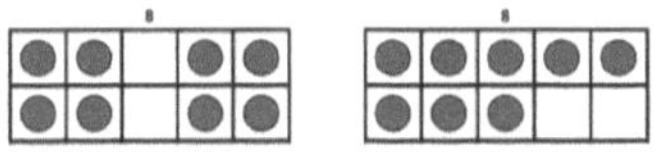

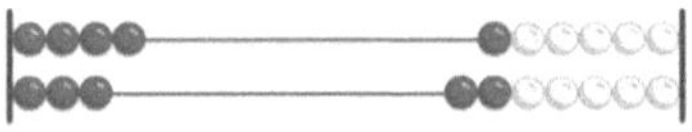

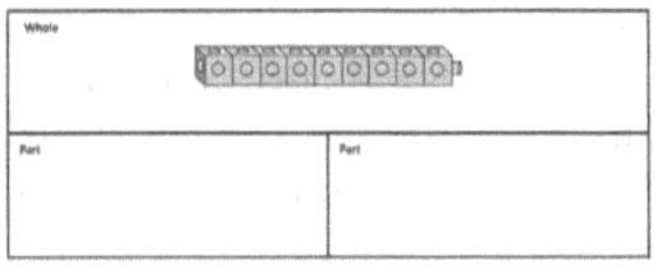

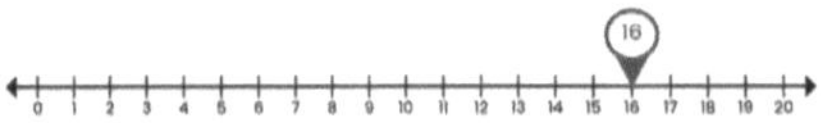

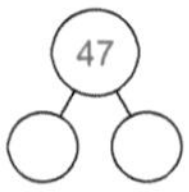

Available for download at **https://online qrs.ly/psf6a5o**

ASSESSING SUBITIZING AND DECOMPOSING

Assessing subitizing and decomposing requires noticing student thinking as they look at quantities (e.g., dots or manipulatives), as they describe how many they see and how they see it, and as they explore physical and mental ways to decompose quantities.

SUBITIZING AND DECOMPOSING LOOK-FORS

Subitizing means seeing quantities without counting. Thus, observing how students determine how many items they see is essential! Students need to be able to decompose flexibly and endlessly. It might seem that your students can decompose numbers flexibly because they break a number apart in one or two different ways. For example, they might use place value and something else that's rather simple, such as one more and one less (e.g., 28 broken into 27 and 1). Without digging deeper, you might overlook that those are the only two ways the student can decompose. Thus, you want to be sure that your students can

- perceptually subitize quantities of 5 or less,
- conceptually subitize quantities within 10 by seeing familiar parts (i.e., seeing 7 as 6 and 1),
- decompose numbers by place value (tens and ones), and
- decompose numbers in many different ways.

QUICK ASSESSMENTS FOR SUBITIZING AND DECOMPOSING

The prompts in Activity 2.5, along with the routines, games, and centers, are good opportunities to gather evidence of your students' progress. The tasks below show some other options. You can do these in small groups or in a one-on-one setting. They should not take more than a few minutes.

Show a face of a die and ask, "How many dots do you see?" Repeat with dot patterns of five or less that are not in a die pattern (e.g., stickers on paper plates).	Have students make a stick of linking cubes (total between 12 and 20). Ask them to break the stick apart in different ways and use equations to describe how they broke it apart.	Share the numbers 83, 73, 63, and 53. Break 83 into 4 and 79. Ask the student if they agree. Then, ask the student to break the other numbers into 4 and something. Observe how they do it.
Quickly flash a subitized amount (e.g., 4). Ask students what they see or how many they see. Then, flash another amount that adds on to the first amount. Ask the students how they determined the amount in the second flash.	Show a number such as 53. Decompose it into three parts (40, 10, and 3). Pose a new number to students and ask them to decompose it into three parts. Watch to see if they replicate your pattern. If so, ask them to do it a different way.	Pose one or more numbers. Ask students to break apart each number in three different ways.

EXPLICIT INSTRUCTION FOR SUBITIZING AND DECOMPOSING

Explicit instruction for subitizing involves more experience with seeing small quantities in a variety of ways. With experience, students come to see quantities. With decomposing, explicit instruction brings attention to the flexible ways in which a quantity or number can be decomposed into parts. In an intervention setting, it can be useful to show different ways to decompose a number (e.g., 8) and then explicitly attend to which of those options are the most useful for a given problem (e.g., 9 + 8 or 28 + 8).

 # ACTIVITY 2.1
DOMINOES SUBITIZE AND SORT

Students develop subitizing skills over time. Therefore, repeated exposure to quantities in different configurations is beneficial for children developing subitizing skills. The main thing to keep in mind is to encourage students to share how many pips or dots they see without counting. This activity focuses on conceptual subitizing—seeing a quantity by subitizing subsets of that quantity. Dominoes are an effective manipulative for such subitizing.

Make sure to have enough dominoes for each student (or pair of students) to sort (about 10, depending on age/experience). Give the dominoes to students and let them engage in an open sort. A prompt might be, "Sort these dominoes in any way you would like." Provide time for students to do their sort. If students are not sure how to sort, offer a suggestion: "Can you sort your dominoes into two groups: four or more dots (pips) and less than four?" Ask students to sort in other ways, as time allows.

Your follow-up conversation could be simply showing a domino and asking students how many dots (pips) they see. Having students explain their thinking is valuable. A student might say, "I see two dots and four dots, so 2 + 4 is 6" or "I see 2 and 2 and 2." Ask students to point to the domino as they are sharing their thinking, as this will help other students see where the numbers are coming from. This is particularly helpful in intervention settings, as you can determine if a student is able to see a quantity of 4 (subitize up to 4). If not, keep sorting and using dominoes (or a die).

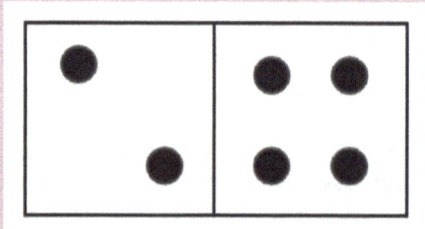

To increase the complexity of the task, discuss possible options for sorting. In an intervention, you can sort and ask the student to determine what your sorting rule was. In the classroom, choose the work of a student or group of students to display (e.g., under a document camera or on the carpet with students sitting in a circle). Ask students, "What do you think is the mystery rule?" Invite students who think they know the answer to place a new domino in the group it belongs in. Other students can agree or disagree with how that domino is placed.

This activity can easily become a center activity. Students could sort the dominoes in any way they choose. They could find all the dominoes that show a specific number. A more sophisticated set of dominoes could be used. The possibilities are endless!

ACTIVITY 2.2
PARTS AND WHOLE MODELS

One may think of a number in parts or those parts put together. The number 67 could be considered as 50 + 17, 30 + 37, or 62 + 5. The more students engage in breaking numbers apart, the more comfortable they will become seeing numbers as both a whole and combination of parts. Linking cubes are a good tool for students to use, as they can easily be taken apart and put back together. Counters, such as teddy bear counters, are also easily moved to show parts of the whole. Explicitly connecting concrete tools to numbers is particularly important in an intervention setting to support student reasoning and assess the understandings that the student has developed.

TEACHING TAKEAWAY

Include contexts to help students reason about different ways to decompose (e.g., number of blocks to get to school or number of bears in the river and on the land).

Students need about twenty linking cubes in one color and a part–part–whole placemat. Ask students to build a tower with a certain number of linking cubes and place the tower in the *whole* section of the mat. Then they can break the tower apart and move the two parts to the corresponding sections of the parts and whole work mat, as shown below. Notice that the students create the whole and place it in the top row. Then, they break it apart and move it to the *part* sections in the bottom row.

Whole		Whole	
Part	Part	Part	Part

Whole		Whole	
Part	Part	Part	Part

Whole		Whole	
Part	Part	Part	Part

Once students have decomposed the number, invite them to share their thinking with a partner. Ask them to verbalize their break-apart (by saying, "I broke 9 into 5 and 4," for example). Ask students to record equations to match their action. The equations for this example could be 9 = 1 + 8, 9 = 6 + 3, and 9 = 5 + 4 (and others). If students are using teddy bear counters, they can engage in the same way, and you can assign one part as on the river and one as on land.

If students are working in a classroom setting, have them work with partners and build the same whole. After they break apart the whole, have them compare how they each did it ("What is alike and what is different?"). In a debrief conversation with the whole class, ask, "Did you break apart the number in the same way or different than your partner did?" You can challenge students to find all the ways to decompose the number. This might be done for a few numbers, but don't overdo it, as it might take away from students manipulating the cubes themselves. For larger numbers, place-value blocks can be used. For flexible decomposing, this will require trading out tens for 10 ones.

The placemat can be downloaded at **https://qrs.ly/psf6a5o**

ACTIVITY 2.3
BREAK IT, MOVE IT, PROVE IT

A number bond is a way to represent decomposing and an effective recording technique to transition to writing equations. It is beneficial for students to have ample practice opportunities for breaking numbers apart, as this helps them see the relationships between the parts and the whole. Breaking apart physical materials first and then representing the action on the number bond will contribute to a deepened understanding of this important mathematical concept.

This activity is described as an interactive classroom activity, but it is also effective in a one-on-one intervention setting wherein the instructor and student exchange their work.

Each student needs a set of about sixty cubes, a sticky note, and a "Break It, Move It, Prove It" recording sheet. Give the students a range of numbers to work within; for example, you might say "Pick a number that is between 40 and 50." Students will write their choice within this range at the top of a number bond and build the number they wrote on their number bond with cubes. Now it is time to "Break It." Each student breaks their cubes into two piles and writes the number of cubes in each part (pile) on the corresponding section of a number bond. Next, students place their sticky note over the total to hide it. They are now ready to challenge their peers (or their teacher). The students will keep the cubes as they are (broken apart) and leave them this way for the next student. When all students are ready (i.e., their work looks like this visual, if they are using place-value blocks), indicate it is time for students to "Move it! Prove it!" by saying this phrase.

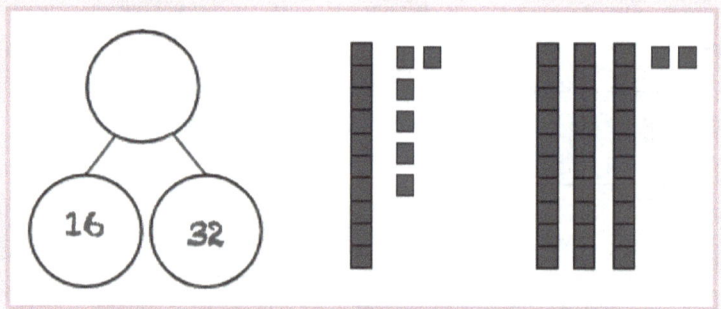

This is a student's cue to move to another student's seat. At this new location, they will "Prove It": figuring out the whole using the blocks and/or the numbers. When the students have an answer, they can lift the sticky note to see if their answer for the whole is correct. At this point, students can rotate to a new seat and repeat or return to their own station and start from the beginning, selecting and building a new number bond.

In a debriefing conversation, ask students to share what strategies they used to prove that the two parts equaled the whole. Invite them to talk about other things they learned about breaking numbers apart. You could extend this activity into a station or partner work, as students can never get enough practice decomposing and composing numbers.

Break It, Move It, Prove It

Directions: Students write a number at the top of the number bond. Then they break it into parts, writing those numbers at the bottom of the number bond. The student hides the top number with a sticky note. Another student "proves it" by determining the hidden number and checks under the sticky note to see if they are correct.

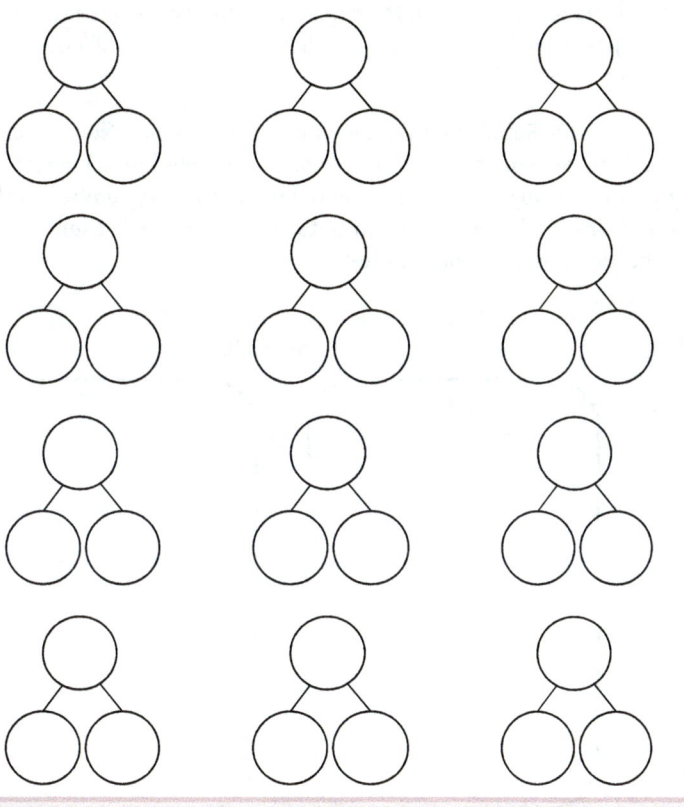

online resources ➤ This resource can be downloaded at **https://qrs.ly/psf6a5o**

ACTIVITY 2.4
EXPRESS IT

Students benefit from thinking about how numbers can be decomposed in various ways, as this supports their emerging number sense and flexibility. This activity encourages students to think of different ways to decompose two-digit numbers by place value, which becomes useful as students work with adding and subtracting multi-digit numbers. For example, when subtracting 73 – 36, it would be useful to consider 73 as 6 tens and 13 ones.

Before students engage in this activity, hang up small posters around the classroom. The posters can be half-sheets of paper or large sticky notes. Students will be rotating from poster to poster, so make sure there are enough posters for students to spread out. On each poster, write a two-digit number. Each student will need a recording sheet, a clipboard or other sturdy surface to write on, and a writing utensil.

Distribute students among the posters. Set a timer for the amount of time you think would work best. As the activity is first rolled out (or if your students are younger), more time may be needed for each rotation. Once the timer begins, students individually come up with as many ways as they can to break apart the number on the poster. They record their thinking using "sticks and dots." The *sticks* represent ten dots and the *dots* represent ones. When the timer sounds, students move clockwise to the next poster in the rotation.

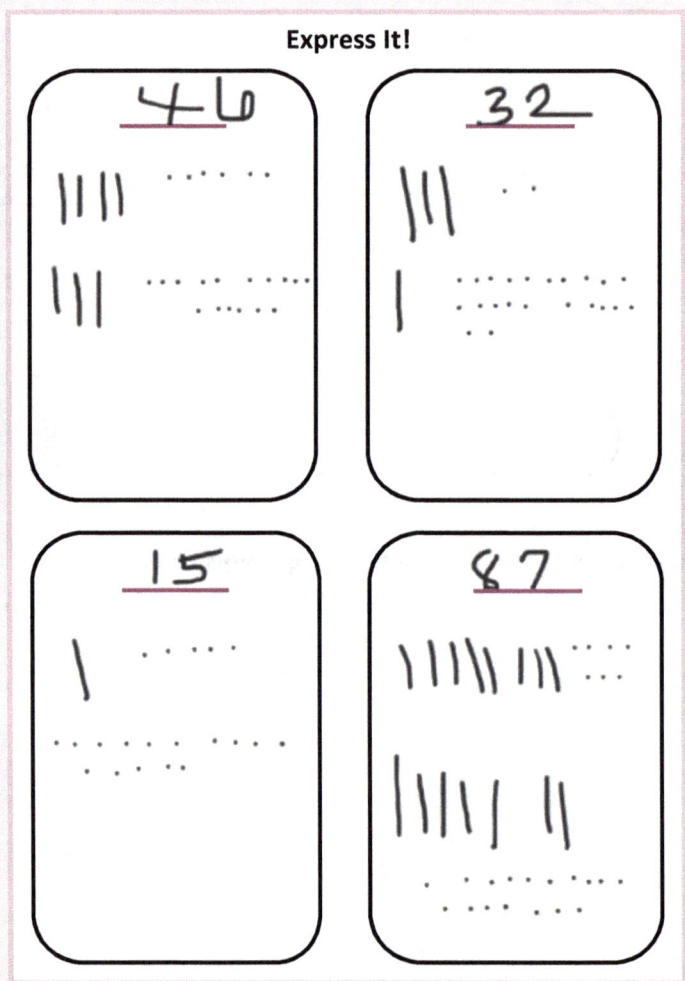

A blank sheet for this resource can be downloaded at **https://qrs.ly/psf6a5o**

Another way this activity could be used is for students to record their decompositions by writing expressions (e.g., for 55, a student might write 40 + 15.) Whichever way the decompositions are recorded, it is important to have a follow-up conversation as a class. Choose a number from one of the posters and ask students to share the different ways they decomposed the number. Honor the various ways students offer by drawing their ways of decomposition on the board. Organizing the expressions on the board may help students discover connections between the different break-aparts. For example, you might list 40 + 15 near 30 + 25 for 55.

This activity may eventually become a routine. Present a number to students and have them offer different expressions for how the number might be broken apart. An example of this might be to display a number (such as 76) and have students suggest different ways it could be decomposed. Student thinking should be recorded on a public space such as a whiteboard or chart paper.

TEACHING TAKEAWAY

Teaching activities like "Express It" can become quick routines; using the activity as a routine gives students ongoing experience with flexible decomposing.

ACTIVITY 2.5
PROMPTS FOR SUBITIZING AND DECOMPOSING

Use the prompts below as opportunities to develop students' understanding of and reasoning with the strategy. Have students use representations and tools to justify their thinking, including base-ten models, number lines, number charts, and so on. After students work with the prompt(s), bring the class together to exchange ideas. Remember that these could be useful for collecting evidence of student understanding.

- Lindy has five horses and they can come into the barn or go out to the field. What are possible ways the five horses might be in either the barn or the field?

- Jonah and Frankie are playing with some dominoes. Jonah says he sees 3 and 4 dots on a domino, but Frankie says she sees a 3 and 2 twos. Could they be looking at the same domino? Explain your thinking.

- During a quick image activity, Lena said she saw three rows of 5, but one dot was missing. Collin said she saw four groups of 3 and 2 more. Were they describing the same quick image?

- You are looking at a poster that has four rows of three dots. What are different ways you could break the dots into groups to describe how many dots you see?

- A student breaks apart 72 into 60 and 12 and another breaks it apart into 30 and 42. What are some other ways you might be able to break apart 72 into two parts?

- Max decomposes 30 into three parts. What might the parts be? Find as many different combinations as you can.

- Poppy is trying to decide how to spend the next 60 minutes. If she wants to do two different activities, how might Poppy spend her time? How long might she do each activity?

- It is ___ miles to your destination. Since it will be a long trip, you decide to travel for more than one day. How might you do this? Which option do you feel is best? Explain your reasoning.

- Sarah wonders if a number can be broken into more than two numbers. What would you tell her? Show your thinking.

QUALITY PRACTICE FOR SUBITIZING AND DECOMPOSING

Traditionally, mathematical topics are contained within the given unit of study, with occasional review. Subitizing and decomposing require ongoing meaningful experiences until students can see small quantities and can readily decompose numbers within 10 automatically. Decomposing grows to lots of other numbers—larger whole numbers, fractions, and decimals. These foundations for fluency are critical and must be revisited, reinforced, extended, and enriched throughout the year. So, if you're working in a measurement unit, it's a good idea to mix in subitizing and decomposing practice. If you're teaching geometry, it's OK to start the lesson with a distance from 10 and 100 routine.

ACTIVITY 2.6

Name: *"Quick Images in Color"*   **Type:** *Routine*

About the Routine: A quick image routine flashes a quantity and has students talk about the amount they think they saw. It often features arrangements so that students can subitize the amount. This routine builds on that by flashing images consecutively so that students can see relationships, number combinations, compositions, and decompositions. Quick images can be made in different ways. An online template with a few examples is provided. Another way to make quick image cards uses paper plates and different colored, 1-inch dot stickers. The example below the directions gives you an idea of what this might look like.

Materials: A set of quick images

Directions:
1. Flash a quick image card to students. Show the image for approximately two seconds.

2. Ask students to tell a partner how many dots they think they saw.

3. Have the class share what they think they saw. Then, show the image again to confirm the amount.

4. Flash a new quick image card that is directly related to the previous image.

5. Have students talk about how many dots they saw the second time and how they know how many they saw.

6. Confirm the amount and record an equation that represents the amount the students saw in the second image.

In these examples, the teacher used paper plates and dot stickers to make their quick image sets. The number 4 is flashed first. Students confirm that they saw four dots. Then, the second plate is flashed with the same four dots and three more. You would want students to say they saw 7 on the second plate because it is 3 more than 4. The two on the right show how you can create quick images with different arrangements and complexities: First, 8 (composed of 5 and 3) is flashed. Then 5 more is flashed, creating 13. The equation for the second could be written as 8 + 5 = 13. Of course, there are other ways to record when more complex examples are used. For the first, the equation 5 + 3 = 8 might have been used. Then, 5 + 5 + 3 might have been used.

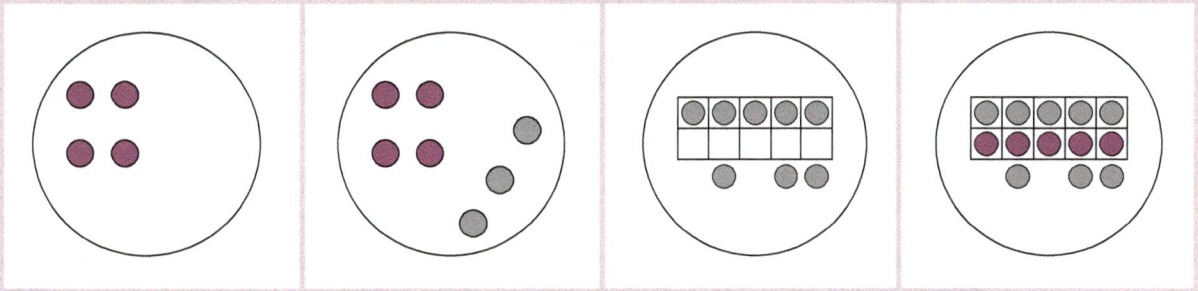

A premade set of quick images to get you started can be downloaded at **https://qrs.ly/psf6a5o**

ACTIVITY 2.7

Name: "The Big Red Ten"　　　　　　　　**Type:** Routine

About the Routine: Subitizing is usually done with dots arranged in regular and irregular patterns. You can intentionally sequence the quantities you flash to students to build on ideas of number composition and relationship as described in Activity 2.1.

This routine builds on subitizing and connects to place value. In it, you present small black dots that represent ones and big red dots that represent tens (this is how the routine gets its name). The idea is that students can subitize 4, and if they see four big reds, that is 40. If they see four big reds and three small black dots, then the image is 43.

As mentioned in the Teaching Takeaway with the last activity, the red ten might be problematic for students who are colorblind. You can change out the large red circle with a large blank (unfilled) circle with a bold border and 10 written in the center. Keep in mind that students need to understand that the big red dot represents 10 small black dots.

Materials: Collection of dot arrangements for subitizing

Directions:　1. Present a dot image. When you flash this image, give students approximately three seconds to see the amount.

　　　　　　　2. Briefly talk about how many dots students saw.

　　　　　　　3. Present the next image that has the same number and arrangement of black dots and one to four big red tens.

　　　　　　　4. Briefly talk about how many dots the students saw this time.

　　　　　　　5. Present a third image that has the same number of black dots and a different number of big red tens.

　　　　　　　6. Talk again with the class about how many dots they saw.

Example:

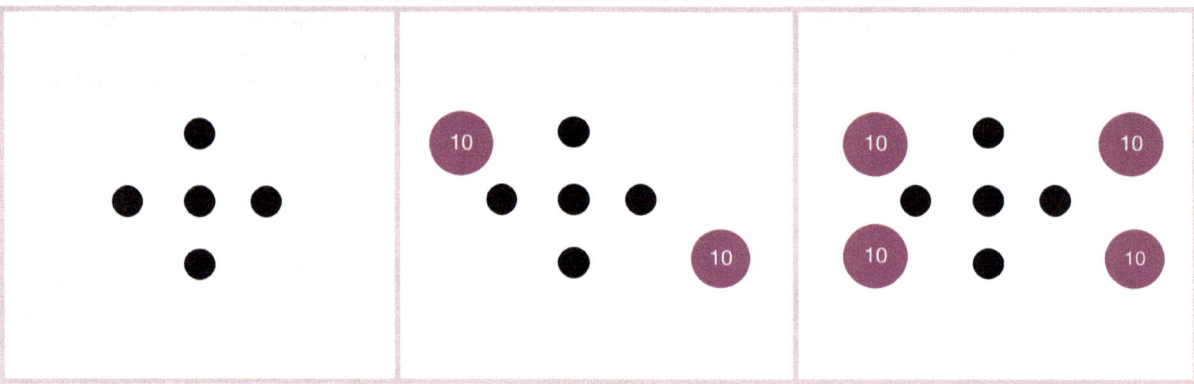

In the example, students might see 5 in the left example as 3 and 2 or merely as 5. Then, when the next image is flashed, they should see 5 and 2 tens, making 25. Then, in the last image, they may see the 25 from the middle example and 2 tens more (45) or they may see four big red tens (40) and five black dots (to make 45).

online resources　Sample dot arrangement slides can be downloaded at **https://qrs.ly/psf6a5o**

ACTIVITY 2.8

Name: *Finders Sitters*   **Type:** *Game*

About the Game: *Finders Sitters* is a fun little game that gets your students up and moving while they practice breaking apart numbers. The goal of the game is for students to find a partner who has decomposed a number into the same part. When they find a match, they sit down. The goal is to find a partner before a timer goes off or before music stops playing. At first, let students decompose however they want. Know that they'll likely stick with obvious decompositions because it's likely that others will have the same parts (e.g., 68 as 60 and 8 or 67 and 1). When this happens, begin to give them conditions for their decompositions, such as "One number must have five tens" (e.g., 68 as 58 and 10 or 55 and 13), "Both numbers must be two-digit numbers" (e.g., 68 as 53 and 15, 44 and 24, or 32 and 36), or "Neither number can be a 10" (e.g., 68 as 58 and 10 wouldn't be allowed).

Materials: Sticky notes, scrap paper, or a whiteboard

Directions: 1. Pose a number to students (e.g., 53).

2. Have them decompose the number into two parts in three different ways (e.g., 40 and 13, 52 and 1, and 26 and 27) and record each way (on a sticky note or scrap paper).

3. Have students stand up.

4. Start a timer or play music to signal students to walk around to find a classmate who has decomposed the number in the same way that they have. Students with matching decompositions sit down.

5. When the music stops or the timer runs out, solicit some of the students' ways to decompose the number and record them on the whiteboard.

6. After recording four or five options, have all students stand and repeat the activity, excluding the options recorded on the whiteboard.

ACTIVITY 2.9

Name: Hot Number Potato   **Type:** Game

About the Game: *Hot Number Potato* is a take on the traditional game of Hot Potato. The difference in this game is that the "potato" is a number written on an index card or small whiteboard. Players can only pass the card (or whiteboard) after they record a way to decompose the given number. Decompositions cannot be repeated. The player left holding the potato gets eliminated until only one player is left. To begin, limit decomposing to two addends; later, students can generate decompositions using up to three addends.

Materials: Index card or small whiteboard and marker, timer

Directions: 1. Put students into groups of four and have them stand.

2. Post or say a number for players to decompose. Have them write it on their index card.

3. When you say "Begin," the player holding the number writes and says aloud one way to break it apart and passes it to the next player.

4. The next player has to record and say a different way to decompose the number. If that way has already been stated by another player, they have to come up with a new way before they can pass the index card to the next player.

5. Players pass the index card as soon as they record a new idea.

6. This is repeated until time has run out. Then the player holding the index card sits down and is eliminated from the game.

> **TEACHING TAKEAWAY**
> Games can almost always be adapted for other content. For example, *Hot Number Potato* could be played with listing multiples or factors of a number.

Continue playing until only one player is left standing in a group.

There are a number of ways to modify this game. You can have larger groups or you can assign a student in each group to be a recorder and judge to make sure that decompositions aren't repeated.

ACTIVITY 2.10

Name: *Five Ways, Most Ways*   **Type:** *Game*

About the Game: To develop skill with flexible decomposition, students need opportunities to practice decomposing numbers in a variety of ways. *Five Ways, Most Ways* is a whole-class game to do just that. In it, students decompose numbers in different ways and then compare with partners to determine how many unique decompositions their team made. The team with the most unique combinations wins.

Materials: Whiteboard, mathematics journal, or scrap paper

Directions: 1. Put students into teams of three or four.

2. Pose a number for teams to decompose.

3. Have each student decompose the number in five different ways.

4. Students on the same team compare their five decompositions.

5. Players cross off or cancel decompositions they have in common. Then, the team counts the number of unique decompositions, earning a point for each unique method.

6. The team with the most points wins the round.

7. After scores are shared, discuss and record examples of different ways the number was decomposed.

For example, second grade students were tasked to break apart the number 65. Leo, Sam, and Jasmine were on a team. Their ideas are listed below.

Leo	Sam	Jasmine
30 and 35	~~50 and 15~~	64 and 1
31 and 34	59 and 6	~~10 and 55~~
~~50 and 15~~	~~55 and 10~~	~~60 and 5~~
~~60 and 5~~	49 and 16	63 and 2
20 and 45	39 and 26	~~50 and 15~~

In this example, each student has 50 and 15, canceling that decomposition. Leo and Jasmine have 60 and 5, so that decomposition is canceled as well. Sam has 55 and 10 and Jasmine has 10 and 55; these are considered the same because the order doesn't matter. After canceling the common decompositions, the team gets eight points by adding Leo's three remaining solutions to Sam's three and to Jasmine's two.

TEACHING TAKEAWAY

As your students become more skilled with flexible decompositions, you can begin to add conditions to the activity. For example, you can tell them to decompose the number into three numbers (e.g., 65 could be thought of as 50, 10, and 5).

ACTIVITY 2.11

Name: Bag of Blocks: Predict and Break   **Type:** Center

About the Center: This center is a way for students to independently practice decomposing numbers with base-ten blocks. Students predict how they can decompose the number into two parts and then use the base-ten blocks to prove their prediction. At first, you should allow students to break apart the blocks however they like. In time, you can provide one of the numbers and charge them with finding the other part. For example, you can give them the starting number of 48 and tell them that one part is 23, causing them to find the other part as 25.

TEACHING TAKEAWAY

Laminate recording sheets or put them in plastic sheet protectors so they can be reused for a center.

Materials: Paper bags (or something similar) with linking cubes or base-ten blocks in them, Bag of Blocks recording sheet (optional)

Directions:

1. A student selects a bag of blocks and empties it.

2. The student finds the number that the blocks represent (e.g., 38).

3. The student predicts how they can break apart the number (e.g., 30 and 8) and then uses the blocks to prove their break-apart is correct. To do this, the student breaks the number apart by separating some of the blocks.

4. The student writes a check mark if their prediction is correct. If it isn't, they put a line through their prediction and write the correct decomposition next to their original prediction.

5. The student repeats this with other bags of blocks.

The following example shows a student who used the recording sheet to organize and show their thinking. Their first bag was Bag A, which had the number 38. They predicted they could break it into 32 and 6. They confirmed their prediction was correct. Then, they predicted they could break 38 into 15 and 12. That was incorrect. They crossed it off their prediction and wrote the correction. You can see their work with Bags D and F as well.

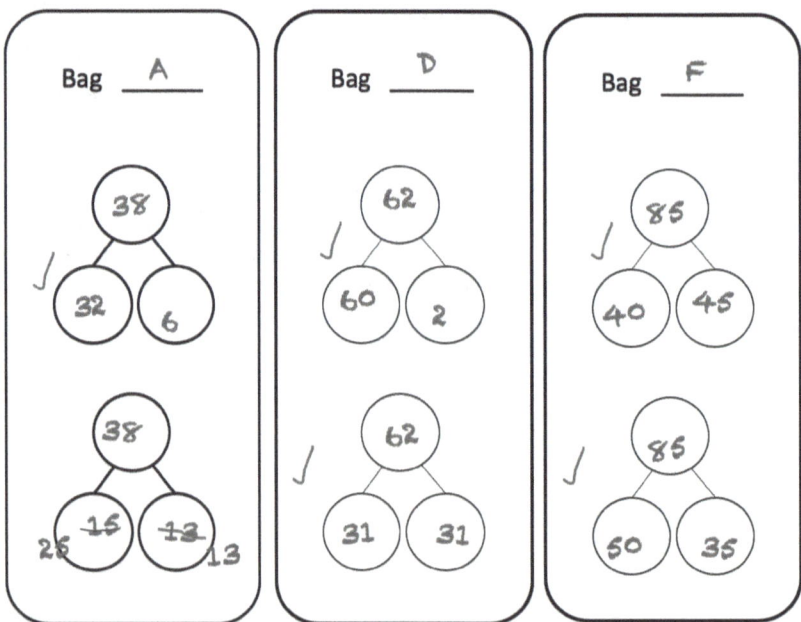

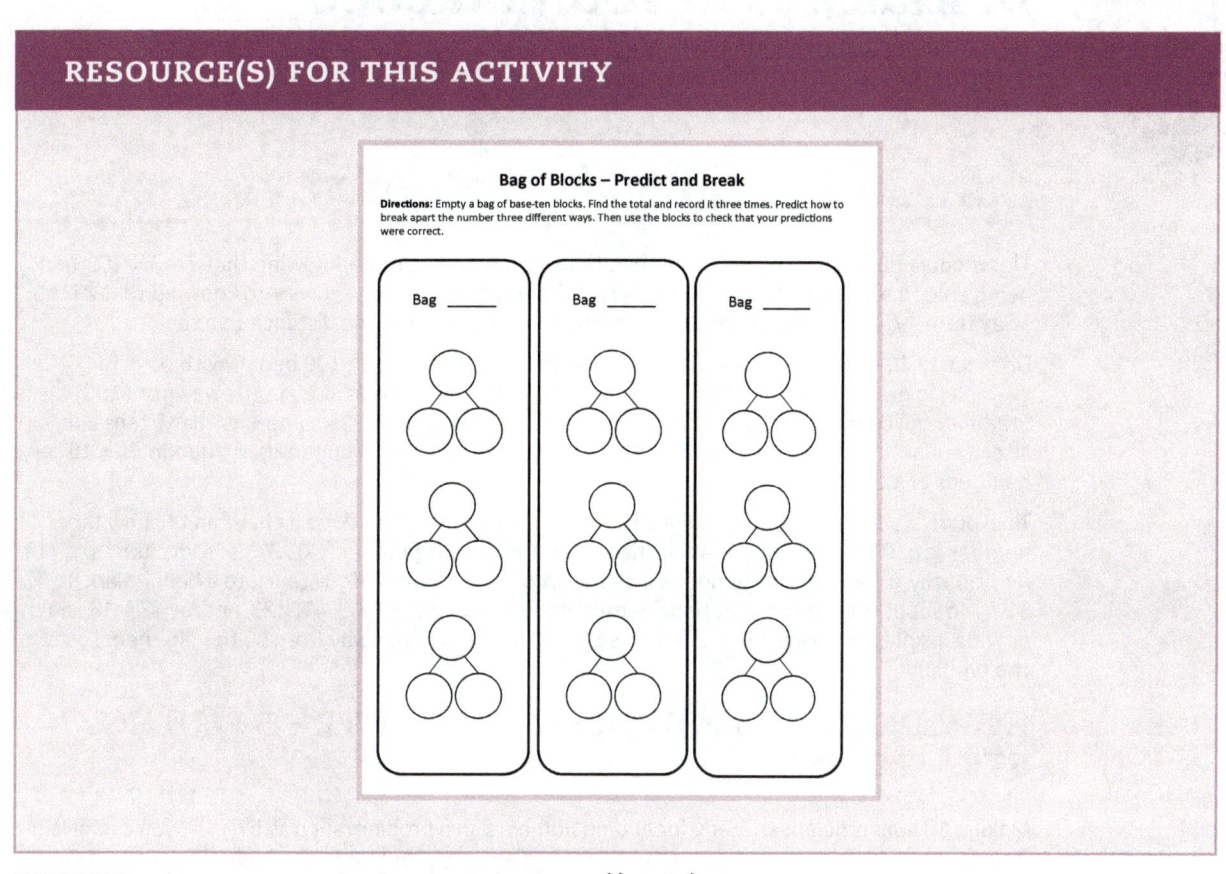

Bag of Blocks – Predict and Break

Directions: Empty a bag of base-ten blocks. Find the total and record it three times. Predict how to break apart the number three different ways. Then use the blocks to check that your predictions were correct.

Bag _____   Bag _____   Bag _____

online resources   This resource can be downloaded at **https://qrs.ly/psf6a5o**

Distance to 10, 100, and 1,000

DISTANCE TO 10, 100, AND 1,000 OVERVIEW

This module builds on the idea of combinations of ten: for example, knowing that 7 + 3 = 10, then being able to see the number 7 and know it is 3 away from 10. This grows into knowing that 27 is 3 away from 30, or 89 is 1 away from 90—in other words, knowing the distance to a ten.

Distance to 10 extends to distance to 100 and to 1,000. Distance to 100 begins with tens; for example, knowing that 60 and 40 combine to equal 100 (which means 60 is 40 away from 100). Students can connect the idea that 6 ones and 4 ones equal 10 in the same way that 6 tens and 4 tens equal 10 tens or 100. And this applies to hundreds as well: 6 hundreds + 4 hundreds = 10 hundreds or 1,000.

To explore student understanding of this foundation, begin with asking if they can tell how far a number (e.g., 7) is from 10; also ask if they can tell how far a number (e.g., 37) is from a ten (e.g., 40). Additionally, it is important to notice if the student uses a number's distance to a benchmark as a way to support their computational reasoning. For example, 47 + 3, 47 + 23, and/or 47 + 16 can provide insights into whether students use the idea that 47 is 3 away from 50 to help them solve the problem.

HOW DOES THIS MODULE CONTRIBUTE TO FLUENCY STRATEGIES?

Making 10 is an efficient strategy for adding numbers. In a problem such as 57 + 27, you can give 3 to either number to make the next ten, creating an easy problem to then add. The student work shows two different ways this strategy can be used for finding the sum.

To use the strategy, the student needs to recognize a combination of 10 (7 and 3) and decompose the other addend accordingly. While either addend can be decomposed and made into a ten, the addend closest to the next ten or hundred is usually the most efficient. For example, in 392 + 516, you could give 4 to 516 to make 520, but it's likely more efficient for you to give 8 to 392, making 400 + 508.

Distance to 10, 100, and 1,000 includes both building toward the next 10 or 100 and thinking about the distance to the previous 10 or 100. While finding the distance to the previous seems less challenging, its application is still very necessary. You can see it in 392 + 516 as you might adjust 516 down to 510. It is more apparent in a problem such as 93 – 56, where a student counts back to solve the problem.

$$93 - 56$$
$$93 - 50 = 43$$
$$43 - 3 = 40$$
$$40 - 3 = 37$$

In this work, the student breaks 56 apart. They count back 50 from 93 to 43. Then, instead of counting back by ones, they break the remaining 6 into 3 to get to 40 and 3 left to count back, arriving at the difference of 37.

The distance to 10, 100, and 1,000 reference page provides an overview, important representations, connections to fluency, and actions to avoid. It can be downloaded and used for reference in planning, teaching, and discussions with colleagues and families.

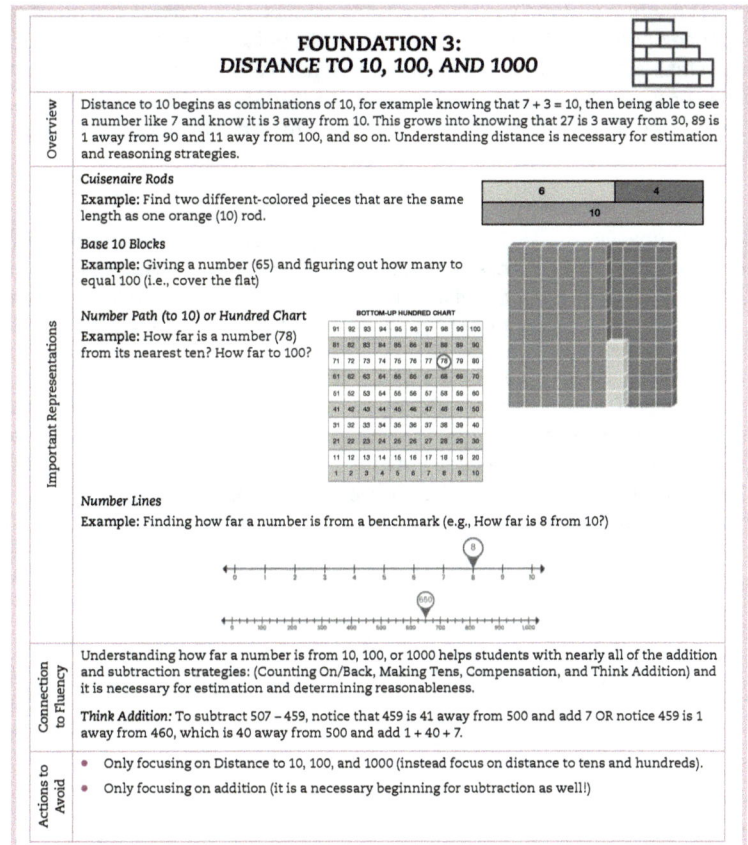

online resources ▶ Available for download at **https://online qrs.ly/psf6a5o**

ASSESSING DISTANCE TO 10, 100, AND 1,000

Assessing this foundation includes observing students as they determine how far away benchmark numbers are. This may involve using number lines, hundred charts, or grouping manipulatives. Additionally, you may want to assess students' mental reasoning without using a physical representation.

DISTANCE TO 10, 100, AND 1,000 LOOK-FORS

You want your students to be able to readily recognize the distance from a number to the next 10, 100, or 1,000. This would include 4 and 6 make 10, 24 and 6 make 30, and even 14 and 86 make 100 (though not as necessary yet). Be on the lookout for students who appear to do this but who are simply counting quickly. As in the problem 93 – 56, when they get to 43 – 6, they don't find 40 and then 37 but rather count back quickly by ones. Though this isn't a major issue in younger grades, it can have lingering effects as students move through the grades. You should find that your students can

- make combinations of 10, 100, and 1,000;

- tell the distance to the next 10, 100, or 1,000 from a given number; and

- easily identify the distance to the previous 10 or 100.

QUICK ASSESSMENTS FOR DISTANCE TO 10, 100, AND 1,000

The prompts in Activity 3.5, along with the routines, games, and centers, are good opportunities to gather evidence of your students' progress. The tasks below give some other options. You can do these in small groups or in a one-on-one setting. They should not take more than a few minutes. We have provided options; select a couple to start with and go to a third if the results are inconclusive or if later observations indicate your student(s) might need additional work with the concept or skill.

Ask a student to show you the different ways they can break apart 10.	Tell a student that you can break apart 10 into 7 and 3 or 6 and 4. Ask them how they might break apart 100.	Start with 33 + 7 = 40. Ask students how they can use that information to determine 53 + 7, 73 + 7, and 93 + 7.
Give students a number and ask them how far it is from the previous ten and the next ten. For example, pose the number 23 and ask students how far it is from 20 and how far it is to 30. Repeat with other numbers.	Use 10-sided dice or digit cards. Randomly make a number. Have the student tell you how far it is to the next and/or previous ten or hundred.	Create a three-digit number. Ask students to tell you how far it is from the next/ previous ten and hundred (e.g., how far is 246 from 250, 240, 300, and 200).

EXPLICIT INSTRUCTION FOR THE DISTANCE TO 10, 100, AND 1,000

To teach distance to (and from) 10, 100, and 1,000, along with combinations of tens, hundreds, and thousands, focus on patterns. The instructional activities in this module will help you think about how to do that. You'll see that using ten frames with place-value disks is one way to do this, but there are many more. Be careful to avoid teaching shortcuts that set students up for problems down the road. For example, you don't want them to think of 30 + 40 as "3 + 4 and then add a 0."

ACTIVITY 3.1
TEN FRAMES AND COUNTERS

Ten frames are a standard tool for learning about and working with numbers. This instructional activity isn't a lesson but rather a variety of ideas for you to build a lesson from. Each uses ten frames to help students make ten, but notice that the counters used in the frames are different. First, two-colored counters are used. This can help students see the combinations as well as patterns within those combinations. Typically, each counter represents one. But here, a black permanent marker has been used to write a 1, 10, or 100 on both sides of the counters as an alternative to place-value disks. Remember that either (counters or disks) are non-proportional representations of numbers (Van de Walle et al., 2023) and students must first understand that a counter with a 10 on it represents ten ones. Again, the ideas below are merely starting points. Create your own investigations or build on activities you have done in the past.

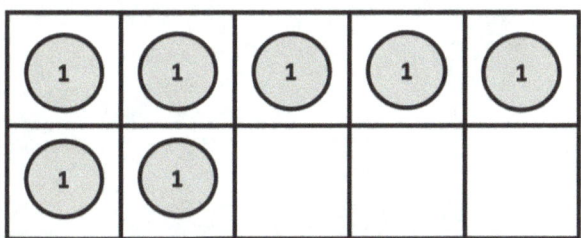

Have students build examples of a number; in the example, 7 is shown. Then, have them determine how many more are needed to make 10. Have them write an equation that matches their thinking. To make combinations of 100, use counters with tens on them.

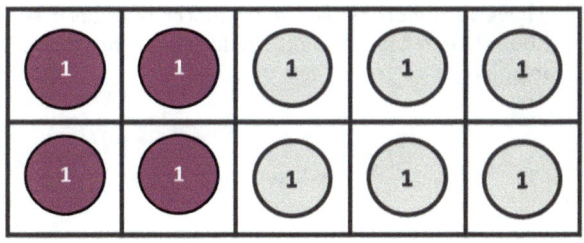

Have students make multiple examples of combinations of ten using both sides of the counters. Here are two examples of 6 and 4 make 10. This is an important experience for students because some may believe that 6 and 4 must always be arranged in a certain way. After making examples, have students explicitly connect each to the same equations that show the combination. As noted above, use counters with tens for combinations of 100 and hundreds for combinations of 1,000.

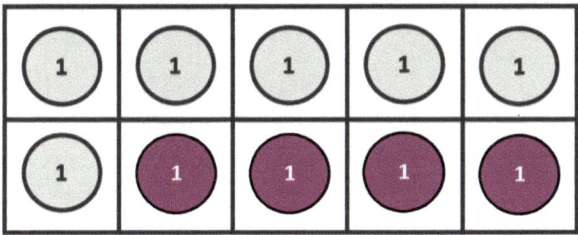

(Continued)

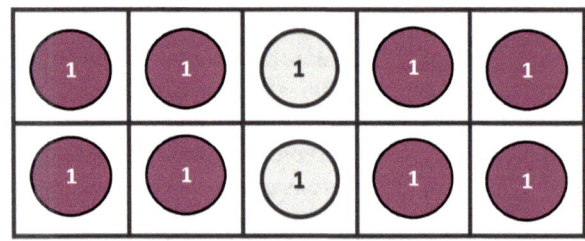

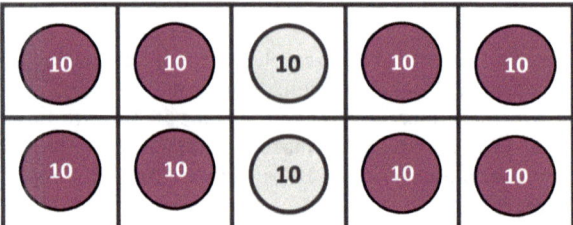

Have students build a combination of 10 using 8 and 2 as shown in the first ten frame. Then, have students build a similar model with counters of ten as shown in the second ten frame. Have students show how the two models are related and how knowing a combination of ten can help with a combination of 100. In certain grades, a third example with counters that have 100 on them could be created to show combinations of 1,000.

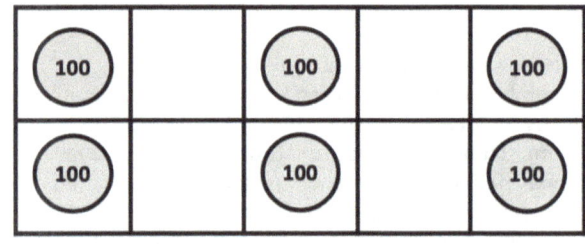

This shows how students could work with combinations of 1,000. You could say to students, "I have six hundreds. How many more hundreds make 1,000?" Then have students build a model and write an equation to prove their thinking.

 ## ACTIVITY 3.2
CUISENAIRE TENS AND HUNDREDS

Cuisenaire rods are good tools for developing a conceptual understanding of fractions. They're also good for teaching multiplicative comparison and even combinations of ten and one hundred. This activity can cement student understanding and recognition of combinations of ten and one hundred. Before reading further, it is important to know that Cuisenaire rods are measured in centimeters (image on the left). There are ten rods, with the orange rod being 10 centimeters and each subsequent rod being 1 centimeter less. Also know that there are 10 millimeters in a centimeter (see the image on the right).

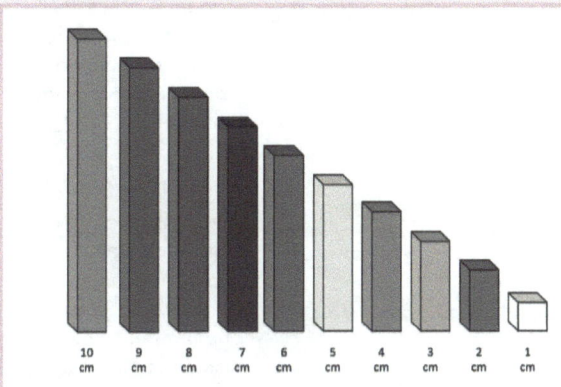
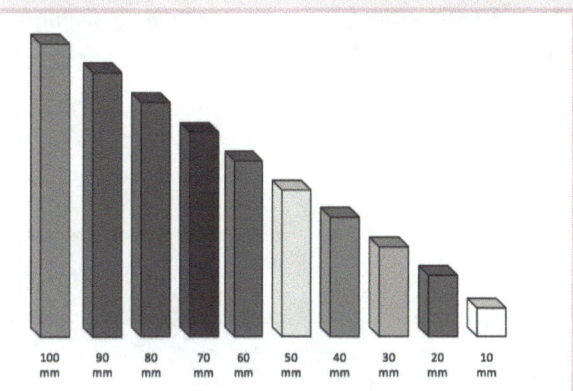

First establish the length of each Cuisenaire rod with students. Have students explore different combinations of rods that equal the length of one orange rod (10 centimeters). They'll likely start with combinations of two rods, as shown in the left column of the table. Have students work to record the equations that their combinations represent and prove that their recordings are accurate.

RODS	EQUATIONS
White + Blue	9 + 1 = 10
Blue + White	1 + 9 = 10
Brown + Red	8 + 2 = 10
Light Green + Black	3 + 7 = 10

To focus on distance to a ten, place a Cuisenaire rod (e.g., the 6 rod) above an orange rod and ask students, "What piece do I need to make 10?" Give students time to think and then ask for their answers; invite a student to place the piece to see if it indeed completes the distance to ten. Repeat this activity using two orange rods (20). Show a quantity (e.g., 15) and ask, "What piece to I need to make 20?"

After working with tens, extend to combinations of 100. This is where the beauty of Cuisenaire rods and the metric system come into play. First, have students use rulers to determine how many millimeters long each rod is. Once millimeter lengths are determined for the white rod and a few others, discuss the relationships of the length of the rods in the two different units (centimeters and millimeters). Be sure students understand that the orange rod is 100 millimeters long. Once confirmed, repeat the exploration above, but this time by recording equations in millimeters. Possible findings appear in the table.

RODS	EQUATIONS
White + Blue	90 + 10 = 100
Blue + White	10 + 90 = 100
Brown + Red	80 + 20 = 100
Light Green + Black	30 + 70 = 100

ACTIVITY 3.3
ARE WE THERE YET?

A context most children understand is driving somewhere and wanting to know how much further it is to the destination. That context is the inspiration for this activity. Place a number line under the document camera, in front of a student, or in the middle of a circle of students. Explain that the 0 represents home and the 10 represents school (pick whatever destinations you want). Ask, "How much further does Joanie have to walk to get to school?" Move "Joanie" to a new place on the number line. Change out the number line for longer ones (e.g., 20, 100). After some experiences, transition to merely saying a value and having students visualize the number line in their heads.

TEACHING TAKEAWAY

Adding a context to a problem makes it more concrete for students and contexts can be connected to students' interests. Rather than walking, Joanie could be riding a bike, running, riding a horse, etc.

TEACHING TAKEAWAY

The context of traveling a distance is an excellent connection to developing a strong understanding of distance to a 10, 20, 100, and so on.

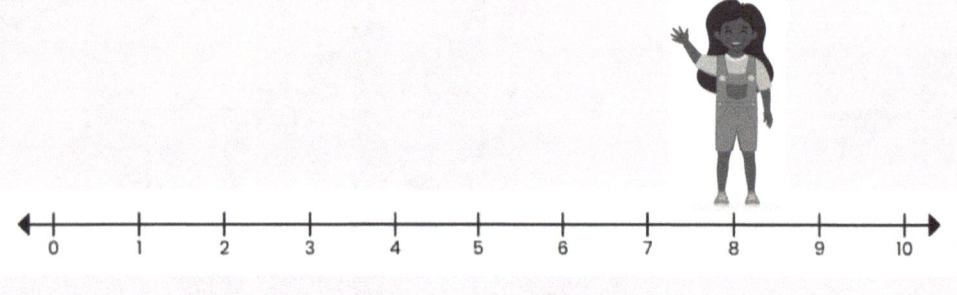

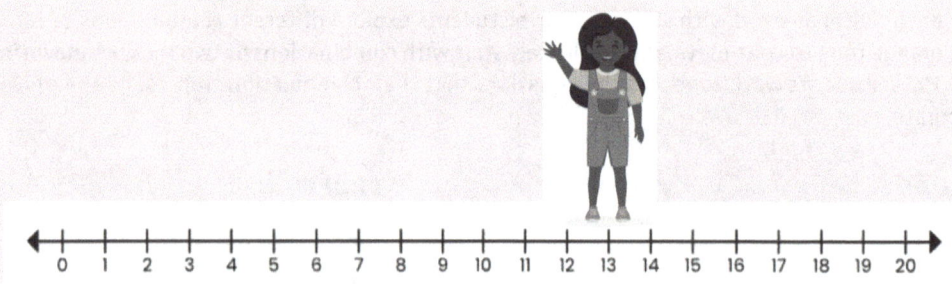

Icon source: iStock.com/Alina Kotliar

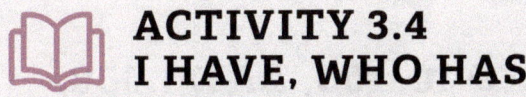

ACTIVITY 3.4
I HAVE, WHO HAS

This activity is a fun way to get students engaged in finding distances and combinations. Write two-digit numbers such as 43, 58, and 17 on index cards and give them out to students. Then, give other students cards with single-digit numbers such as 7, 2, and 3. Have students stand up and mingle with classmates in search of a partner who they can combine with to make the next ten. Examples are listed in the table below.

Partner A: 43	Partner A: 58	Partner A: 17
Partner B: 7	Partner B: 2	Partner B: 3
43 and 7 make 50	58 and 2 make 60	17 and 3 make 20

After students find a match, have them work on a whiteboard to justify that they are compatible. They might record equations, draw base-ten blocks, or do something else. Once matches have been made and proven correct, bring the class together to record combinations on the whiteboard and ask students to share the patterns they notice in those combinations. One option with this activity is to have some cards without a match. Once you have recorded matches and talked about patterns, you can then have students without a match share what number they needed.

To add a bit of novelty to the activity, you can choose to give out blank cards or sticky notes to students instead. When doing this, tell half of the class to record a two-digit number and the other half to record a single-digit number. Then, carry out the activity in the same way. In time, you can have students form groups in a variety of ways, including those listed in the table below.

MAKE ANY 10	MAKE ANY 100
Partner A: 38	Partner A: 380
Partner B: 42	Partner B: 420
They match because 38 and 42 makes 80.	They match because 380 and 420 makes 800.

ACTIVITY 3.5
PROMPTS FOR DISTANCE TO 10, 100, AND 1,000

Use the prompts below as opportunities to develop understanding of and reasoning with the strategy. Have students use representations and tools to justify their thinking, including base-ten models, number lines, number charts, and so on. After students work with the prompt(s), bring the class together to exchange ideas. Remember that these could be useful for collecting evidence of student understanding.

- You want to buy ten cookies at the cookie shop. When you get there, you notice that there are two types of cookies, oatmeal and sugar. What are the different combinations of cookies you might buy?

- Latrenda says she can make 100 in five different ways. How do you think she might do it?

- Two different numbers make 100. What are three possible sets of numbers? Choose one of your sets and prove that it does make 100.

- Cody wants to run 10 kilometers. When he passes the 7-kilometer mark, how much further does he need to run?

- Lindy is collecting stickers from every park she visits. Her goal is 100 stickers. If she has 78, how many more does she need?

- Luka says that because 2 + 8 = 10, he knows that 20 + 80 = 100 and 200 + 800 = 1,000. Is Luka's thinking on track? Share why you think it is or isn't.

- The golf team has a goal of hitting 100 balls during their practice. Each person hits a multiple of ten balls. If there are four people on the team, what might be the number of balls they hit to make their goal?

- Gigi is working on the problem 57 + 38. She adds 3 to 57 and gets 60. Then she adds 35 to 60 and arrives at the answer of 95. Do you think Gigi's strategy works? Explain why or why not.

- Create a double-digit subtraction problem and share how you might use a ten to solve it.

- Carver says, "I like to get to a decade number when I add." What might Carver mean when he says this? Can you help explain his strategy to a friend for the problem 47 + 29?

- How is making 10 similar to making 100? How is it different?

QUALITY PRACTICE FOR DISTANCE TO 10, 100, AND 1,000

When done well, practice can unlock patterns and relationships for students. It can position them to see how an idea applies to different numbers. It helps them become efficient and clever. For example, making ten (4 + 6) connects with making 70 (64 + 6). Use lessons and games to help students see and make connections. Modify a game that practices making ten to make 20, 30, or 80. Help students see how combinations of ten are related to combinations of 100, then have them practice those relationships.

ACTIVITY 3.6

Name: "Complex Number Strings (Combinations of 10, 100, and 1,000)"   **Type:** Routine

About the Routine: A number string is a list of equations that highlight a pattern within sums and differences, products and sums, and so on. A complex number string is a matrix of equations that presents patterns both vertically and horizontally. See the example below to see how this is presented. They are fairly easy to create. A complex number string is perfect for helping students see patterns within distances or combinations of 10, 100, and 1,000. Seeing the patterns helps many students generalize these combinations so that they can apply them to all sorts of situations.

TEACHING TAKEAWAY

Complex number strings are easy to make with a table in a slide deck that can be projected on a screen. Alternatively, you can make a poster and laminate it to use repeatedly.

Materials: This routine does not require any materials.

Directions: 1. Provide a matrix of related expressions with one known sum. It makes sense for this to be a basic combination, such as 8 + 2 = 10 (Combination of 10 example) or 60 + ___ = 100 (Distance to 100 example).

2. Give the class a choice of working down the first column or across the first row.

3. After students signal that they know the sums of a row or column, hold a class discussion about how each known solution relates to the others. Draw students' attention to how knowing a basic combination of 10 (e.g., 8 + 2) can be used to find other sums. After completing and recording the products for an entire row or column (determined by the class), have students complete the remaining cells independently or with partners.

Distance to 10 Example

As you discuss patterns, point out how the sum changes in relation to the change in addends. For example, 18 + 2 is 10 more than 8 + 2 because 18 is 10 more than 8.

8 + 2 =	8 + 12 =	8 + 22 =	8 + 32 =
18 + 2 =	18 + 12 =	18 + 22 =	18 + 32 =
28 + 2 =	28 + 12 =	28 + 22 =	28 + 32 =
38 + 2 =	38 + 12 =	38 + 22 =	38 + 32 =
48 + 2 =	48 + 12 =	48 + 22 =	48 + 32 =
58 + 2 =	58 + 12 =	58 + 22 =	58 + 32 =
68 + 2 =	68 + 12 =	68 + 22 =	68 + 32 =

(Continued)

(*Continued*)

Figuring Out Fluency—Addition and Subtraction With Whole Numbers

Distance to 10, 100, and 1,000 Example

Distance examples include a missing value, with the answer being 10, 100, or 1,000. As students work through the strings, they use patterns and relationships to find their sums. For example, they might compare the missing values that equal 100 and equal 1,000 and how they used that relationship to find the missing value.

9 + __ = 10	8 + __ = 10	7 + __ = 10	6 + __ = 10
60 + __ = 100	80 + __ = 100	30 + __ = 100	70 + __ = 100
69 + ___ = 100	88 + ___ = 100	37 + __ = 100	76 + ___ = 100
690 + ___ = 1,000	880 + ___ = 1,000	370 + ___ = 1,000	760 + ___ = 1,000

Remember, the foundations in these modules build toward larger numbers and different number types. An interventionist working with decimals might first strengthen students' understanding and skill related to distance to 10 and 100 before moving to distance to 1,000 or distance to 1. Your students' skill with building 10, 100, 1,000, or 1.0 will grow. When it does, you can consider shifting the experience to Using Combinations of 10, 100, 1,000, or 1.0 as shown.

Using Combinations of 1,000				Using Combinations of 1.0			
6 + 4 =	6 + 5 =	6 + 6 =	6 + 7 =	6 + 4 =	6 + 5 =	6 + 6 =	6 + 7 =
600 + 400 =	600 + 500 =	600 + 600 =	600 + 700 =	0.6 + 0.4 =	0.6 + 0.5 =	0.6 + 0.6 =	0.6 + 0.7 =
600 + 450 =	600 + 550 =	600 + 650 =	600 + 750 =	0.6 + 0.45 =	0.6 + 0.55 =	0.6 + 0.65 =	0.6 + 0.75 =
600 + 460 =	600 + 560 =	600 + 660 =	600 + 760 =	0.6 + 0.46 =	0.6 + 0.56 =	0.6 + 0.66 =	0.6 + 0.76 =
600 + 470 =	600 + 570 =	600 + 670 =	600 + 770 =	0.6 + 0.47 =	0.6 + 0.57 =	0.6 + 0.67 =	0.6 + 0.77 =
600 + 480 =	600 + 580 =	600 + 680 =	600 + 780 =	0.6 + 0.48 =	0.6 + 0.58 =	0.6 + 0.68 =	0.6 + 0.78 =
600 + 490 =	600 + 590 =	600 + 690 =	600 + 790 =	0.6 + 0.49 =	0.6 + 0.59 =	0.6 + 0.69 =	0.6 + 0.79 =

ACTIVITY 3.7

Name: "How Many to Ten? With Quick Images"

Type: Routine

About the Routine: It is beneficial for students to have opportunities to think about combinations of ten. This routine will encourage students to do this. Students can also apply their knowledge of making a ten to making a hundred and a thousand. They will get practice subitizing with this routine as well by getting quick glances at combinations of numbers to determine the value of each set.

Materials: Digital How Many to Ten? With Quick Images slides or premade cards with different arrangement of place-value disks; three colors of transparent chips, circle counters, or die-cut circles to put values on for center use (nine of each color); permanent marker

Directions: 1. Flash an image to students. Only make the image visible for a short time—approximately two seconds—as you want students to subitize, not count the dots. Students indicate they have an answer of how many dots by putting their thumb up close to their chest.

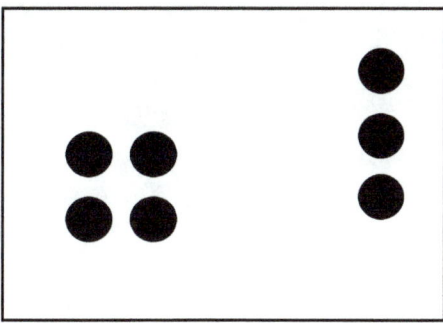

2. Briefly have students share the ways they saw the dots and how many more are needed to make ten. A student might say, "I see 7 because 4 and 3 make 7. Three more are needed to get to 10."

3. Repeat the routine using cards with different dot arrangements.

At first, use black dots with quantities of less than ten. This will get students to think about how many more are needed to get to ten. Eventually, disks with ten and one hundred can be added in.

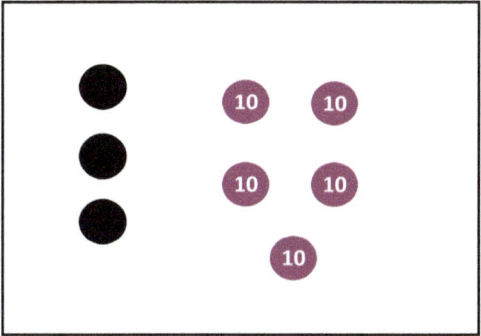
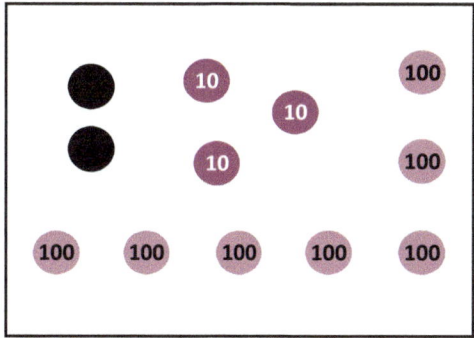

It is important to note that when more disks with larger quantities are added, you need to give students adequate time to look at the dots and get a sense of what is there. But don't give them enough time to count each disk one by one! Responses from students might sound like this for cards that include groupings of ten and hundred disks: "I saw five tens and 3, so 50 and 3 is 53. Seven more are needed to get to the next ten or 60" or "I saw seven hundreds, three tens, and 2. So 732. I would need 8 to get to 740 and 68 to get to 800."

(Continued)

(Continued)

Once students become comfortable with this routine, it can be incorporated into a center. This center could be managed in two ways. The first would be to have students replicate the routine as it was implemented in the classroom. One partner would flash a premade card and the other would determine the quantity and amount needed to get to ten. Then the roles would reverse.

Another way this could be used in a center is to have students create the arrangements with the disks. Partners work together, with one partner making a configuration of place-value disks behind a barrier, such as an open file folder or book. The partner would remove the barrier while the other student would determine what quantity they saw. Then the second partner would say how many would be needed to get to the next ten or hundred. Partner 1, who made the arrangement, would confirm if the answer was correct or ask the student to revise their thinking.

A premade set of images to get you started can be downloaded at **https://qrs.ly/psf6a5o**

ACTIVITY 3.8

Name: Next Ten, Last Ten   **Type:** Game

About the Game: Identifying the next ten is applied with the Count On, Count Back, and Make Tens strategies. For example, when solving 39 + 27, you think about 1 more to 40 or 3 more to 30. Identifying the last ten can help students subtract. *Next Ten, Last Ten* is a game for practicing distance to the nearest ten. The intention here is for students to play so often that they begin to instantly determine the distance to the next ten or how far a given number is from the previous ten. The example below features numbers in the forties. The online site has this version and a blank version so that you can have students practice finding the distances from a variety of numbers.

Materials: Playing cards (queens = 0, aces = 1; remove tens, kings, and jacks) or a 10-sided die; one *Next Ten, Last Ten* game board per player; and two-colored counters or centimeter cubes for game pieces

Directions:
1. Players place ten game pieces on any of the spaces on their game board. Note that a player can place more than one piece on any given space (e.g., three counters on the number 46).

2. Then, players take turns rolling a number and thinking about which number on the game board it combines with to create the next ten or the last ten.

3. If a game piece is on a space with a matching number, the player removes their game piece and takes another turn.

4. If a player can remove a number on three consecutive turns, they can place a game piece on their opponent's game board.

5. If a player cannot remove a game piece, it is their opponent's turn.

6. The first player to remove all of their game pieces wins.

> **TEACHING TAKEAWAY**
> This game can be played frequently to develop automaticity. Automaticity with finding the next ten or the last ten is critical for using significant strategies for addition and subtraction (Count On, Count Back, Make Tens, and Compensation).

(Continued)

(Continued)

The example below shows how the game is played. Player 1 puts 10 counters on their game board: two counters on 42 and on 46 on the Distance to the Next Ten side of the board and two counters on 49 on the Distance to the Last Ten side of the board. On their first turn, Player 1 rolled a 6; 44 is 6 to the next ten (50) but they don't have a counter on that space. The 46 space is 6 from the last ten (40) but they don't have a counter on that space either, so they lose their turn. On their next turn, they roll a 2; 48 is 2 from the next ten, so they remove a counter from 48, as shown in the middle image. They get to go again and roll a 9; 49 is 9 from the last ten, so once again, they get to remove a counter. This is two removals in a row. If they get a third removal on their next turn, they can add a piece to their opponent's game board.

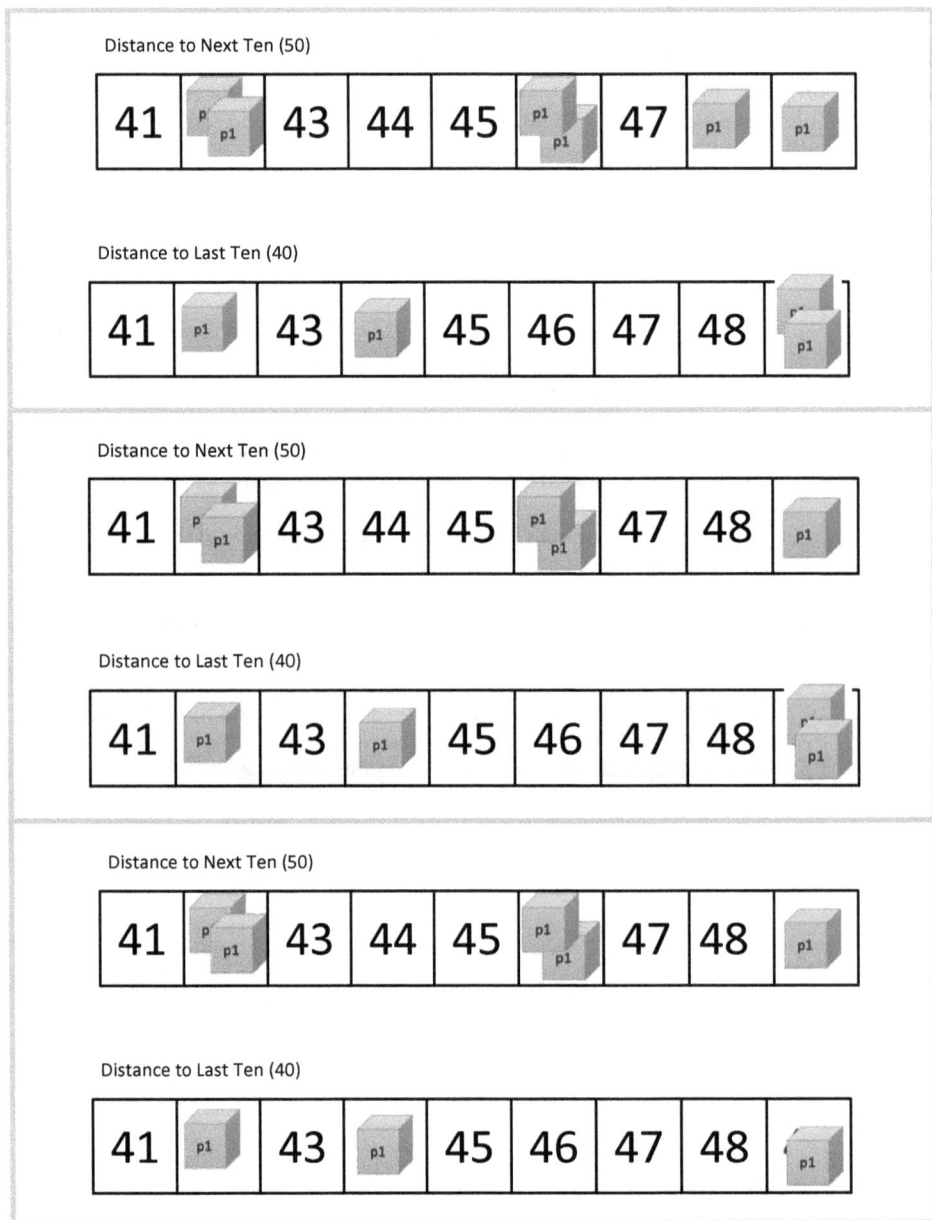

Game boards for this activity can be downloaded at **https://qrs.ly/psf6a5o**

ACTIVITY 3.9

Name: Combinations   **Type:** Game

About the Game: *Combinations* is a game for practicing combinations of ten. Players take turns trying to make combinations of ten. The goal is to be the first player to clear all of the cards they are playing. After students demonstrate proficiency (even automaticity) with Making 10, you can modify the target from combinations of ten to combinations of 12, 15, or some other number. This adjustment reinforces the need for flexible decomposition.

Materials: One deck of playing cards (queens = 0, aces = 1; remove kings and jacks) or four sets of digit cards (0–9) per player

Directions: 1. Each player has their own deck of cards and begins the game by dealing themselves a set of six cards face up.

2. Player 1 looks for cards that combine to make 10 (e.g., 7 and 3). The player tries to make as many combinations as possible. Each card used (e.g., 7 and 3) is cleared from the set of six.

3. When no more combinations can be made, the player's turn is over. At the end of the turn, a player can choose to discard a card or not to discard. Note that discarding could provide a strategic advantage but it also provides an opportunity for the game to end when there are three cards in a player's hand and two are used to make a ten. The third is then discarded to end the game.

4. After Player 2's turn, Player 1 adds a card to their playing set and attempts to make combinations. When no combinations can be made, Player 1 can discard a card or not.

5. Play continues until a player clears their entire hand.

(Continued)

(Continued)

In the example, players are using digit cards. Player 1 begins with 9, 3, 4, 5, 1, and 6, as shown. Player 1 makes combinations of ten with 9 and 1 and 4 and 6. Those cards are cleared (Image 2). Player 1 chooses to not discard (dealing a 2 on their next turn would end the game). On their next turn, they are dealt an 8 (Image 4). They can't make a ten, so they decide not to discard. On their next turn, they are dealt a 5 (Image 5) and can't make a ten. On their next turn, they are dealt a 2 (Image 6). They choose to make a ten with 3, 5, and 2 (Image 7). They choose to discard the 4 to end their turn, leaving 8 for their next turn and hoping to be dealt a 2 to win.

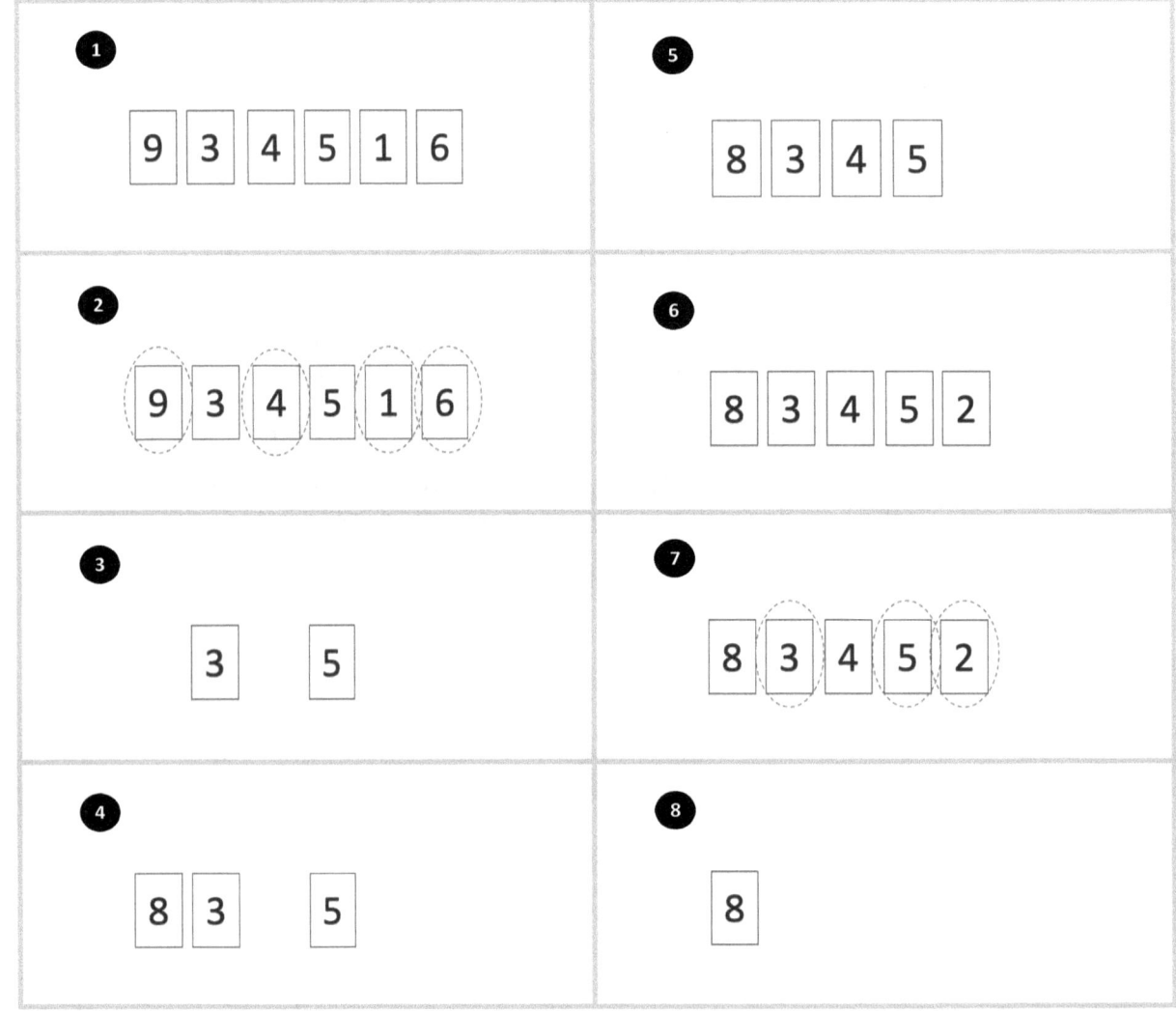

ACTIVITY 3.10

Name: *Parts and Stripes* **Type:** *Game*

About the Game: Distance to 100 is a natural extension of Distance to 10. It supports three-digit whole-number computation and strategies for adding decimals. *Parts and Stripes* is a good game for students to practice finding or recalling combinations of 100.

Materials: Playing cards (queens = 0, aces = 1; remove tens, kings, and jacks) or a 10-sided die; *Parts and Stripes* game board (one per group), centimeter cubes for player pieces (or something similar)

Directions:
1. Players take turns generating a two-digit number. Note that digits can be used in any order: 8 and 4 can be used as 84 or rearranged to be 48. Also note that 0 can be used as the number of ones (e.g., 30) or the number of tens (e.g., 03).

2. The player determines the number that combines with the generated number to make 100. For example, 16 combines with a generated 84 to make 100.

3. The player puts a counter on that number.

4. The first player to get two stripes of four in a row wins the game.

5. Alternative: Once a number is covered, it can't be used again. An alternative is that players can steal spaces from their opponent.

Below, Player 1 flips a 3 and 5, making it 35. They place their counter on 65 because 35 and 65 make 100 (left image). Then, Player 2 covers 37 because they flipped a 3 and a 6 (63), which they combined with 37 to make 100. In the third image, you see that Player 2 has one stripe of four in a row. As mentioned, combinations of 100 help students when they work with decimals. The version of *Parts and Stripes* on the far right shows how it could be played with decimals.

TEACHING TAKEAWAY

When introducing a game with decimals, consider playing a whole-number version of the game first. This can help students strengthen understanding by making connections.

Parts and Stripes

Directions: Players take turns creating a two-digit number. Players determine the number that pairs with the number they generated to make 100. Players put a counter on that number. The first player to get two stripes of four counters in a row wins.

1	2	3	4	5	6	7	8	9	10
11	12	13	14	15	16	17	18	19	20
21	22	23	24	25	26	27	28	29	30
31	32	33	34	35	36	37	38	39	40
41	42	43	44	45	46	47	48	49	50
51	52	53	54	55	56	57	58	59	60
61	62	63	64	P1	66	67	68	69	70
71	72	73	74	75	76	77	78	79	80
81	82	83	84	85	86	87	88	89	90
91	92	93	94	95	96	97	98	99	100

Parts and Stripes

Directions: Players take turns creating a two-digit number. Players determine the number that pairs with the number they generated to make 100. Players put a counter on that number. The first player to get two stripes of four counters in a row wins.

1	2	3	4	5	6	7	8	9	10
11	12	13	14	15	16	17	18	19	20
21	22	23	24	25	26	27	28	29	30
31	32	33	34	35	36	P2	38	39	40
41	42	43	44	45	46	47	48	49	50
51	52	53	54	55	56	57	58	59	60
61	62	63	64	P1	66	67	68	69	70
71	72	73	74	75	76	77	78	79	80
81	82	83	84	85	86	87	88	89	90
91	92	93	94	95	96	97	98	99	100

Parts and Stripes

Directions: Players take turns creating a two-digit number. Players determine the number that pairs with the number they generated to make 100. Players put a counter on that number. The first player to get two stripes of four counters in a row wins.

1	2	3	4	5	6	7	8	9	10
11	P1	P2	14	15	16	17	P1	P1	20
21	22	23	24	25	26	27	28	29	30
31	32	33	34	35	P2	P2	P2	P2	40
41	42	43	44	45	46	47	48	49	50
51	52	P1	P2	55	56	57	58	59	60
61	62	63	P2	P1	66	67	68	69	70
71	72	73	74	75	P2	P1	P1	79	80
P1	82	83	84	85	86	87	88	89	90
91	92	93	94	95	96	97	98	99	100

Parts and Stripes

Directions: Players take turns creating a two-digit decimal number. Players determine the number that pairs with the number they generated to make 1.00. Players put a counter on that number. The first player to get two stripes of four counters in a row wins.

0.01	0.02	0.03	0.04	0.05	0.06	0.07	0.08	0.09	0.20
0.11	0.12	0.13	0.14	0.15	0.16	0.17	0.18	0.19	0.20
0.21	0.22	0.23	0.24	0.25	0.26	0.27	0.28	0.29	0.30
0.31	0.32	0.33	0.34	0.35	0.36	0.37	0.38	0.39	0.40
0.41	0.42	0.43	0.44	0.45	0.46	0.47	0.48	0.49	0.50
0.51	0.52	0.53	0.54	0.55	0.56	0.57	0.58	0.59	0.60
0.61	0.62	0.63	0.64	0.65	0.66	0.67	0.68	0.69	0.70
0.71	0.72	0.73	0.74	0.75	0.76	0.77	0.78	0.79	0.80
0.81	0.82	0.83	0.84	0.85	0.86	0.87	0.88	0.89	0.90
0.91	0.92	0.93	0.94	0.95	0.96	0.97	0.98	0.99	1.00

online resources 🡒 The game boards for this activity can be downloaded at **https://qrs.ly/psf6a5o**

Counting and Skip Counting

COUNTING AND SKIP COUNTING OVERVIEW

Counting and skip counting, both forward and backward, are at the heart of elementary mathematics, beginning in kindergarten with single-digit numbers, then to multi-digit numbers, then counting by ones starting at a number other than 0, and eventually skip counting. But there is so much more to counting (and skip counting)! And these other ways of counting are somewhat hidden in that they are not spelled out directly in some curriculum standards and noted briefly as a sidebar in others. These "hidden" counting standards (SanGiovanni, 2023) or skills include counting back (in first grade), skip counting by an interval from any number, skip counting back, or skip counting by a variety of intervals (twenties, fifties, and so on). Without these experiences, students have difficulty carrying out counting strategies because they rely on counting by single ones or tens and have problems counting back in chunks (e.g., by tens or to a benchmark), which compromises subtraction strategies.

Students' skill at counting or skip counting can be misleading. Students may effortlessly skip count 0, 5, 10, 15, 20 or 2, 4, 6, 8, 10. Yet, they might be reciting a somewhat memorized pattern akin to recalling their ABCs rather than using number relationships. Is this same student able to skip count to solve 79 + 45? If not, they need more counting experiences! That is why it is important for their skip-counting experiences to include starting at any number and skip counting by intervals such as 5 and 10 (e.g., 3, 8, 13, 18, 23, 28, 33 or 6, 16, 26, 36).

This module is provided because counting and skip counting are important yet potentially fragile skills for many students. The instructional activities intend to expose students to a variety of skip-counting intervals and the patterns within skip counts. Most notably, effortless counting and skip counting come with oodles of practice experiences. We have provided a collection of resources to help you provide engaging, frequent practice to ensure students are flexible in skip counting.

HOW DO COUNTING AND SKIP COUNTING CONTRIBUTE TO FLUENCY STRATEGIES?

Counting and skip counting are foundational for the Counting On, Counting Back, and Counting Up strategies. While these three strategies may seem rudimentary, they are both effective and efficient when the counting is done with groups or chunks. For example, finding the sum of 217 + 544 by counting on from 544 by 200, then 10, and then 7 is probably speedier than breaking both numbers apart and finding partial sums. The problem in many of our classrooms is that students struggle to flexibly count forward and backward by chunks.

Additionally, students may do fine with counting forward and backward by ones, tens, and hundreds. But their ability to count by *groups of* ones, tens, and hundreds may still be developing. This skill needs to be practiced so that it can be applied to computation. For example, look at the following comparison: While both students find the sum of 217 + 544 by counting on from 544, the student on the left counts on by single hundreds, tens, and ones whereas the student on the right counts on by chunks of hundreds and ones. Imagine if both students counted on from 217 instead of 544. The student on the left is more likely to make an error and be inefficient as they count on.

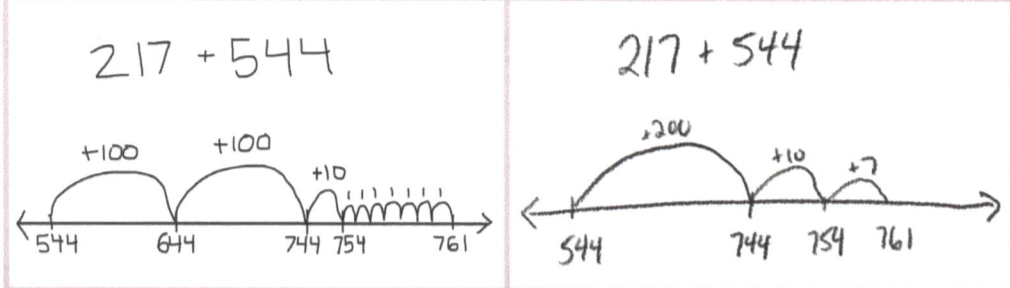

Similar strengths and needs appear when students subtract. In the next examples, both students can skip count backward, which is good! The student on the left, as with the addition example, counts by single hundreds, tens, and ones. The student on the right chunks their counts. Note that they decomposed 80 to count back to a benchmark (200). This is efficient and reasonable but with more experience, the student might be able to make the count from 253 to 173 in one jump.

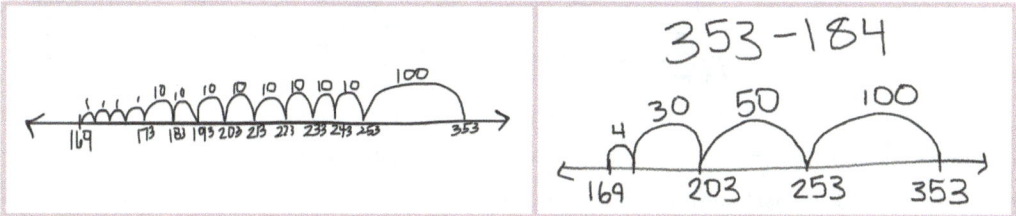

The counting and skip counting reference page provides an overview, important representations, connections to fluency, and actions to avoid. It can be downloaded and used for reference in planning, teaching, and discussions with colleagues and families.

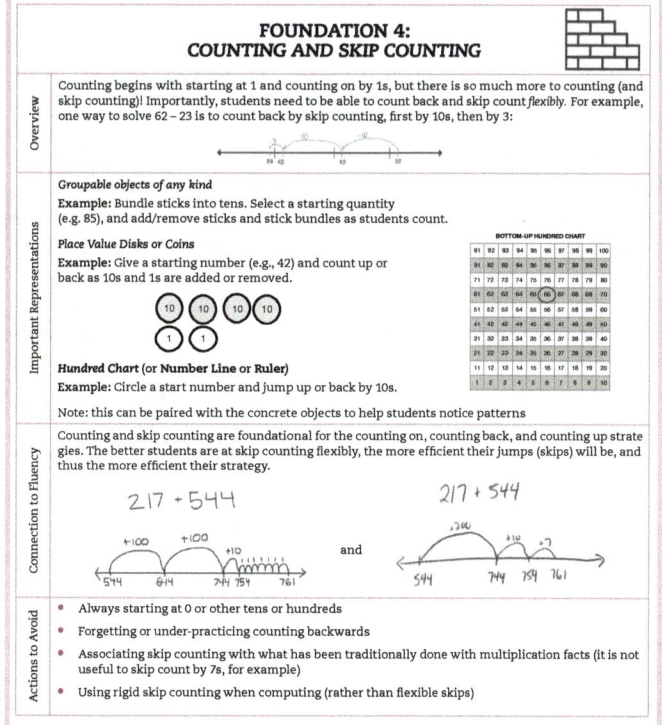

online resources → Available for download at **https://online.qrs.ly/psf6a5o**

ASSESSING COUNTING AND SKIP COUNTING

The goal of assessing student thinking related to counting is noticing to what extent students are becoming efficient and flexible in how they are counting. Here are observation ideas (*look-fors*) and interview or journal prompts (*quick assessments*).

COUNTING AND SKIP COUNTING LOOK-FORS

You first want to be sure that students can count by ones both forward and backward. Be sure that they understand and demonstrate counting principles of stable order, one-to-one correspondence, cardinality, and so on. Look for these same principles to apply to skip counting. Be aware that early skip counters may rely on counting by ones to generate the terms in the skip count. If they are completing a line of skip counts on a piece of paper, their counts or answers may be correct, but their method for reaching the next count may not be a skip count. For example, the student's skip count in the following example seems to show that they can count by tens. Without witnessing it, you can't tell that the student counted, somewhat quickly on their fingers, to each new ten and a little more slowly from one number to the next, especially from 96 to 106.

$$36, 46, 56, 66, 76, 86, 96, 106$$

Important look-fors related to efficient and flexible skip counting include

- count forward and backward from any given number,
- skip count forward and backward by fives, tens, and hundreds from any given number,
- skip count forward and backward by reasonable groups of tens or hundreds,
- move between different skip-counting intervals (e.g., count by tens and then ones), and
- decompose ones, tens, and hundreds to flexibly skip count (i.e., chunk).

QUICK ASSESSMENTS FOR COUNTING AND SKIP COUNTING

The prompts below can be used for diagnostic assessments with an individual child or as writing prompts for all students. Use these options as a menu, selecting prompts that fit your context. Adapt numbers as needed. For more ideas, see Activity 4.5, along with the routines, games, and centers, which all provide good opportunities to gather evidence of your students' progress.

Select a starting number. Ask the student to count backward for you. Have them count for a bit.	Share a number. Ask the student what other numbers they would say if they were to skip count forward (or backward) by 10.	Select a starting number. Have a student skip count forward by hundreds. Then, select a different number and have them do it again. Alternatively, pick a number and ask the student to skip count backward.
Pass out a blank hundred chart. Write any number in an appropriate box. Ask the student to complete the number's row and column orally. You can choose to record the numbers as they count.	Select a starting number. Have the student skip count forward by tens. Then, select a different number and have them do it again. Alternatively, pick a number and have the student skip count backward. Ask the student to tell you what patterns are in the counts (e.g., the ones place doesn't change).	Share a number such as 137 or 415 and ask the student to count back 80 and show their count on a number line (if needed). Look for them to count back by singles, groups (10 then 70), or the entire amount.

EXPLICIT INSTRUCTION FOR COUNTING AND SKIP COUNTING

Early instruction of counting begins with the counting principles of one-to-one correspondence, stable order, cardinality, and so on. Students move from single digits to numbers within 20, 100, and beyond. As students begin to skip count, you will find that some revert to counting by singles to carry out a skip count. When this happens, be sure to confirm their accuracy but also explicitly connect their count by singles to the skip count results. Assuming students understand the idea of skip counting, the work in front of you is to provide ample opportunities to rehearse skip counting.

 # ACTIVITY 4.1
30-SECOND COUNTS

Have you ever wondered about how many smiley faces you can make in a minute? If so, you might not have realized you were sitting on the perfect hook for practicing estimation and counting. In this instructional activity, ask students to predict how many smiley faces each individual can make, how many a group of four can make, and how many the whole class can make in 30 seconds. Of course, before you begin, you have to be clear about what constitutes a smiley face. See below for some ideas.

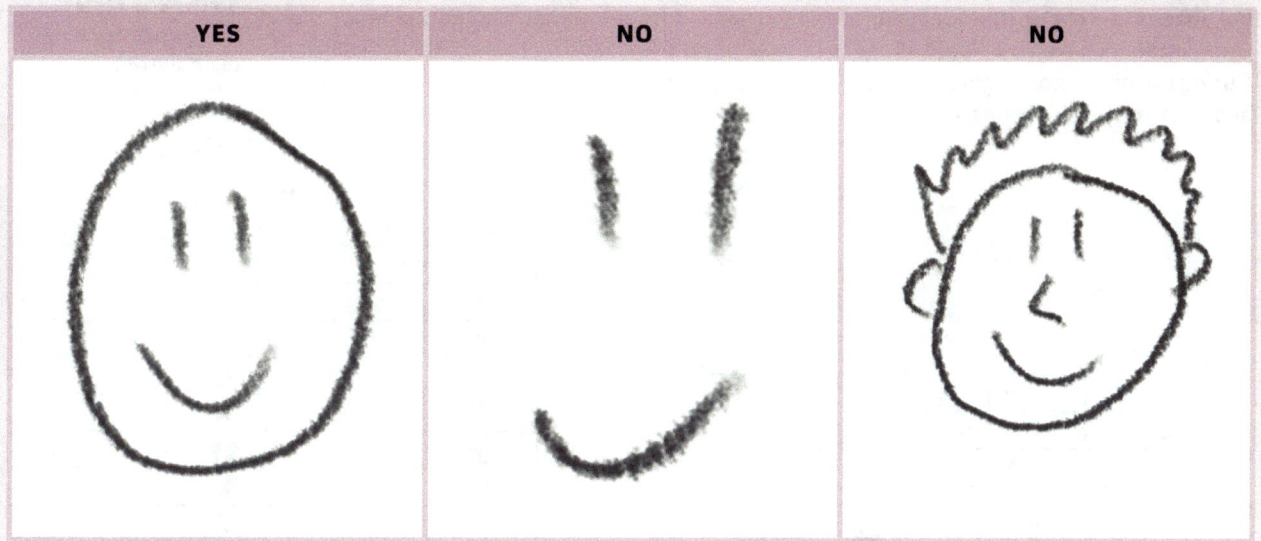

YES	NO	NO

After students make their smiley faces, have them count the amount they made; expect anywhere from 15 to 35 per student. Then ask each group to find the total number for smileys in their group. Challenge the class with figuring out a way to count all of the smiley faces made. Finally, have them compare the actual amount the class made with their predictions. Extend the activity by asking how many they can make in a minute.

Come back to the activity later and ask them to make shapes, icons (e.g., school mascot, hearts, stars, etc.), or letters. Ask them to predict if they can make more of the new shape than the number of smiley faces they made earlier. You could turn this into a skip-counting activity. To do this, replace smiley faces with representations of tens, as in the examples shown below.

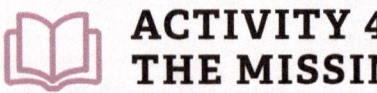

ACTIVITY 4.2
THE MISSING

This instructional activity is a forerunner to a routine of the same name (SanGiovanni, 2020) and would work well as a set of stations that small groups of students rotate through. At each station, you provide a somewhat empty bottom-up hundred chart with some given numbers and some symbols on them, as shown below. Students have to find the number that each symbol represents. Students should talk with each other about how they know their number is correct. You can also use a recording sheet (on the right) for students to write the number for each symbol. After a few minutes, students rotate to the next station to work with a new number chart. Once students have completed all the rotations, bring the class together to discuss solutions. During the discussion, highlight the different ways students counted on and backward by ones and tens or groups of ones and tens.

TEACHING TAKEAWAY

Using a bottom-up number chart helps students communicate mathematically, because going up on the chart is also going up (increasing) quantitatively.

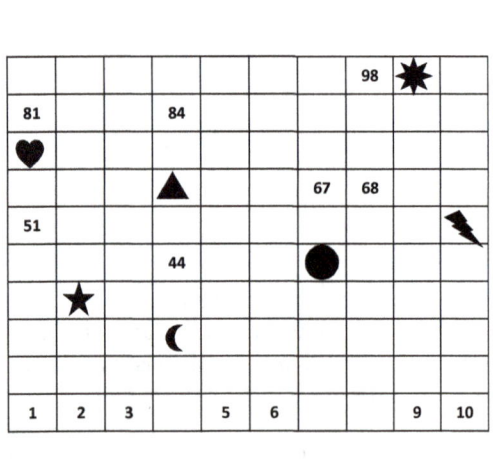

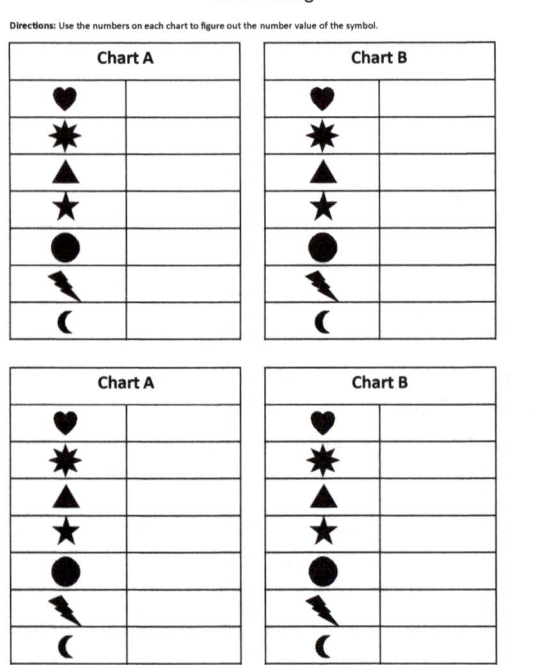

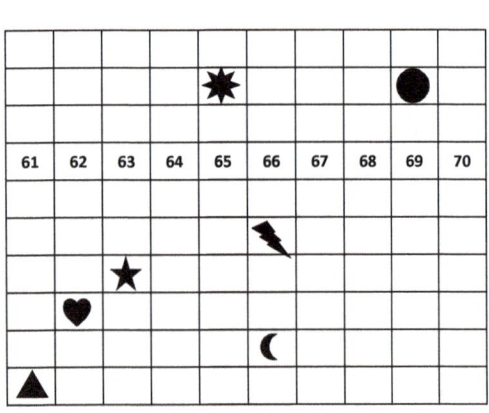

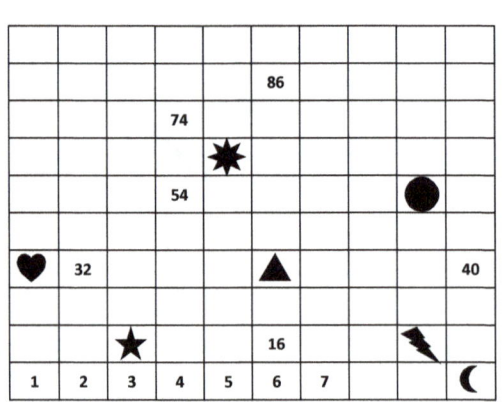

This instructional activity naturally extends to other charts, as shown below. On the left, students can practice counting numbers between 400 and 500. You can even link two or three charts you count within a larger range. On the right, you can see how the activity could be extended for older students who can count well with whole numbers but haven't transferred that skill to decimals.

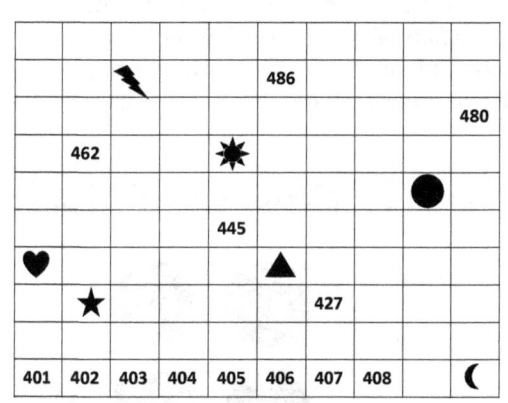 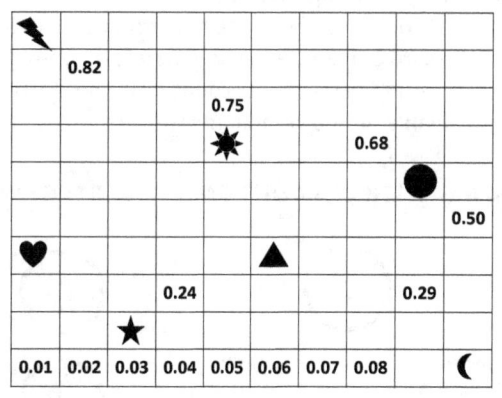

The chart and recording sheets for this resource can be downloaded at **https://qrs.ly/psf6a5o**

ACTIVITY 4.3
DOT CARD COUNTING ON

Counting collections often focus on counting all. Students also need opportunities for counting collections as a counting on or counting back experience. This activity asks students to count on by ones, tens, and hundreds. The arrangements encourage students to chunk their counting on, supporting their flexibility with skip counting. Assign a starting number with a 0–99 card or have students generate a two-digit number with cards or dice to identify the starting number. Then, show a dot card. Have students count from the starting number by the number on the dot card. A large collection of dot cards is available on the companion website.

START WITH 47, COUNT ON	START WITH 32, COUNT ON	START WITH 114, COUNT ON
① 1 ① 1 ① 1 ① 1	10 10 10 10 10 10	100 100 100 100 100 100 100

online resources Dot cards can be downloaded at **https://qrs.ly/psf6a5o**

The downloadable dot cards are designed so that students can count on by ones or by a sub-itized group of tens or hundreds (i.e., chunking their counts). Before counting, have students share how many they see and then reinforce that the starting number and the additional amount (on the card) get you to a certain number. Using the left example, you show 47. You ask how many are on the card and students say 4. Together you count 47, 48, 49, 50, 51. Then, you say 47 and 4 is 51.

TEACHING TAKEAWAY

Record associated equations to help students gain comfort with symbols and make connections between concrete and abstract representations. For example, after counting on from 32, equations might be: 32 + 40 + 20 = 92 and 32 + 60 = 92.

 ## ACTIVITY 4.4
METER STICK COUNT BACK

Meter sticks are excellent number lines that can be used to model counting forward and backward. In this activity, you pose a number such as 82. Students then work with a partner or two to count back the amount you announce (e.g., 27). Your instruction would be, "Start with 82 and count back 27." After you announce it, students count back to identify the number they end with. As student counting proficiencies grow, look for students who count back by singles and students who count back by groups. For example, when counting back 27 from 82, students might count 72, 62, 61, 60, 59, 58, 57, 56, and 55. Call attention to the different approaches to counting, pointing out that the differences yield the same result.

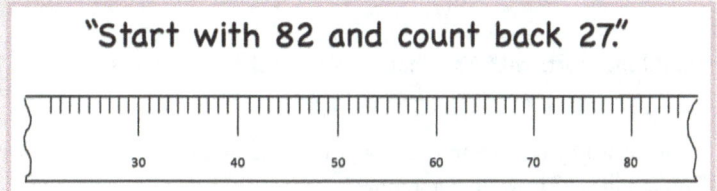

"Start with 82 and count back 27."

You can ask questions before the count to enrich the activity by reinforcing concepts of number and skill with estimation. To do this, pose the starting number and the counting amount as described. Then, ask students to estimate what number they think they'll end with. Record their ideas on the board and compare after counting. Have them cover or flip over their meter sticks so that they can't count before estimating. You can ask other questions: "What other numbers will you say?" "What numbers won't you say?" and so on. As your students show prowess counting within 100, you can put two or even three meter sticks together to count with larger numbers. Of course, to do this, you'll need a larger amount of floor space.

ACTIVITY 4.5
PROMPTS FOR COUNTING AND SKIP COUNTING

Use the prompts below as opportunities to develop understanding and skill with counting and skip counting. Have students use representations and tools to justify their thinking, including base-ten models, number lines, number charts, and so on as necessary. After students work with the prompt(s), bring the class together to exchange ideas. Remember that these could be useful for collecting evidence of student understanding.

- Kara says that when you count, the pattern repeats. Do you think she's counting by ones, tens, or hundreds? Do you think she's counting forward or backward? What pattern do you think she means? Create an example to show your thinking.

- Asia is counting by tens and she starts with 43. What are some other numbers she says? Prove how you would know she would say them.

- Dante started to count backward by tens from 222. He got to 202 and then was stuck. What comes next? What are some other numbers Dante will say? How do you know?

- Pick a number between 21 and 29. Count by ones from that number. Count by tens from that number. Count by hundreds from that number. Record the numbers you say. State how they are different.

- What changes about a number when you count by tens? What stays the same? Give examples to explain your thinking.

- Show two different ways to use skip counting to solve 85 + 65.

- Jax counted back to solve 433 – 150. How might Jax have solved the problem?

- How can skip counting back by tens help you solve 97 – 40? What is another problem that skip counting back can help you with?

QUALITY PRACTICE FOR COUNTING AND SKIP COUNTING

Be vigilant while your students practice. Premature or poor practice can reinforce misconceptions or inefficiencies. Counting and skip counting are good examples of this hazard. For example, students may actually count by ones when they claim to be skip counting by tens or other groups. Counting practice may predominantly feature counting forward. When students do it well, we assume they can count forward and backward. Skip-counting practice often begins with 0. Students rattle off the skip counts without thinking, seemingly showing that they can skip count. Yet, when starting with another number, the skip counting falls apart. Thus, the student will not be able to use this good beginning to skip count from other numbers in order to learn efficient ways to add or subtract (e.g., Count On or Count Back strategies).

ACTIVITY 4.6

Name: "The Count" **Type:** Routine

About the Routine: This quick routine can give you insight into students' proficiency with counting while providing a good opportunity for practice. The Count is a routine that has students estimate and skip count. It is a good opportunity for practicing skip counting by a variety of intervals, which is essential for using the Count On and Count Back strategies. You can extend the routine by ending with a problem connected to the counts. For example, you can have students start with 43 and count on by ones. Then, have them start with 43 and count on by tens. Pose a problem, such as 43 + 48. Have students turn and talk with their classmates about how they can count on by tens and ones to find the sum.

Materials: Identify a counting interval and a starting number, supporting number chart (optional)

Directions:
1. Set a clear counting path so that students know how they will count in the room. Having students gather in a large circle is a good option.

2. Identify a starting number, the student who will count first beginning with that number, and a counting interval such as "skip count forward by tens." Then, ask students about the impending count. Questions to ask include the following:

   - What are some numbers we will say as we count?

   - What number do you think you will say?

   - What will be the last number said?

3. As students count, you can record the numbers they say on the board or mark them on a related number chart or number line. Doing either will help with a post-count discussion about the patterns within the numbers that were said and discussion about the predictions made before the count.

4. After the count, discuss patterns within the numbers said, challenges with counting, and how student predictions compared with the results of the count.

> **TEACHING TAKEAWAY**
> You can increase engagement with routines by turning them into challenges or games. For The Count, you can have students write down the number they think they will say or the number they think the count will end with on a whiteboard. At the end, see how many in the class were accurate and consider a way to give points to the class for their accuracies.

THE COUNT			
Start with 43. Count by ones.	Start with 43. Count by fives.	Start with 43. Count by tens.	Start with 43. Count by twenties.
Start with 43. Count back by ones.	Start with 43. Count back by twos.	Start with 43. Count back by fives.	Start with 43. Count back by tens.
Start with 108. Count by twenties.	Start with 336. Count by fifties.	Start with 517. Count back by tens.	Start with 784. Count back by fours.

This example shows a variety of prompts you can pose. The top two rows show how you can start with the same number and count by different intervals. The bottom row gives some different ideas about starting numbers and counting intervals. Note that counting back is also an important use of the routine (as it supports subtraction) and that there are a variety of intervals provided. Students should have experience working with a variety of skip counts to help them move from counting by singles to counting by groups.

ACTIVITY 4.7

Name: "Stop the Count"   **Type:** Routine

About the Routine: Skill with counting and skip counting clearly supports strategies for Counting On, Counting Back, and Counting Up. When using a strategy flexibly, students don't always skip count by a group of tens and then a group of ones. For example, to solve 37 + 85, a student might skip count by 70 to 107, then 10 to 117, then 3 to 120 and then 2 more to 122. Students might mix skip counting backward and forward. For 142 – 68, a student might count back 70 to 72 and then count on 2 to 74 (Count Back strategy) or they might count up from 68 to 148 (80) and count back 6 (74) (Count Up strategy). This routine is a choral counting opportunity to mix counting forward and backward by singles and groups.

Materials: Signs with directions for counting intervals (see the table below).

1. Announce a starting number for the class count.

2. Show and announce the first counting interval (such as by twos).

3. Have students begin to choral count by the interval from the starting number.

4. Randomly stop the count. Show and announce a new interval (such as by tens).

5. Have the students pick up the count with the new interval from where the count ended.

6. Repeat stopping the count, changing the counting interval, and restarting the count from the last number.

For example, a teacher announces the starting number of 46 and shares the first counting interval of count by twos. The class starts counting, "48, 50, 52, 54, 56, 58, 60, 62." When the class says 62, the teacher exclaims, "Stop the count! Now count by tens." The class picks up the count: "72, 82, 92, 102, 112, 122." The teacher exclaims, "Stop the count! Count by fives." The class picks up the count: "127, 132, 137, 142, 147, 152, 157, 162." The count is changed two more times and the routine ends.

As you work with this routine, you can record the numbers that the class says on your whiteboard. This can be helpful for when errors are made or when the count changes. Recording the numbers can also be useful for pointing out patterns. Have fun with the routine. As students demonstrate flexibility with counting, turn it into a whole-class game where they try to complete a number of interval changes without making errors in the count. Set a goal of counting for a minute or two without making an error. Have fun with the routine!

Each column in the table shows a different counting interval that you might show to your class during the routine. You can write these on copy paper, construction paper, or paper plates. Large printables are available for download.

Count on by ones	Count on by twos	Count on by fours	Count on by fives
Count on by tens	Count on by twenties	Count on by forties	Count on by fifties
Count on by hundreds	Count on by two hundreds	Count on by four hundreds	Count on by five hundreds
Count **back** by ones	Count **back** by twos	Count **back** by fours	Count **back** by fives
Count **back** by tens	Count **back** by twenties	Count **back** by forties	Count **back** by fifties
Count **back** by one hundreds	Count **back** by two hundreds	Count **back** by four hundreds	Count **back** by five hundreds

 Counting interval slides can be downloaded at **https://qrs.ly/psf6a5o**

ACTIVITY 4.8

Name: *Crossing Over*   **Type:** *Game*

About the Game: *Crossing Over* is a game of strategy that reinforces counting forward and backward. It provides an opportunity for students to count and to hear how others count. Through repetition, students can become more and more efficient with their counting. The game is exceptionally adaptable so that you and your students can create all sorts of fun, engaging alternatives to keep the play enjoyable. For example, you can make the starting point 75, 350, or any other number to practice counting from a variety of places. You can change what happens when face cards are played: For example, a king reverses the counting order, but you could choose to make it worth 30 instead. You can change how long the game lasts by having students play with half of the cards instead of the entire deck.

Materials: A deck of playing cards (black [spades/clubs] count forward, red [hearts/diamonds] count backward; aces = 1, jacks = 20, queens = 0; kings = reverse direction of counting), two-colored counters

Directions:

1. To begin play, each player is dealt three cards. The rest of the cards are stacked in a deck face down.

2. The game begins at 50. Player 1 plays a card and counts that amount (black counts forward, red counts backward). When finished counting, the player says the next player's name. Optional: Players who miscount get a chip.

3. Player 1 pulls a card from the deck (players must have three cards in their hand) and Player 2 plays a card to count forward or backward from the number Player 1 ended with. After counting, Player 2 says Player 3's name and draws a card. Player 3 plays a card, counts, and on the cycle repeats.

4. When a player crosses over a decade (forward or backward), they get a chip.

5. When a player ends exactly on a decade (e.g., 30, 40, 50, 60, etc.), they lose a chip.

6. If a player's card takes them to 0 or 100, they get three chips and the count begins again at 50.

7. The player with the fewest chips when all cards are played wins.

The table that follows shows how the game is played. Starting at 50, Player 1 plays a black 8 and counts to 58. Player 2 plays a black 3 and counts to 61. Because they crossed over 60, Player 2 got a chip (black dot). Notice that on Player 2's second turn, they played a red 5, which caused them to cross over 60 again. Now they have two chips. On Player 4's second turn, they played a red jack, crossing two decades (50 and 40) and now have two chips.

(Continued)

ROUND	RESULT AFTER PLAYER PLAYS	PLAYER 1	PLAYER 2	PLAYER 3	PLAYER 4
1	50 (starting number)	Black 8 (counts on 8 from 50)			
1	58 (after Player 1's +8)		Black 3 • (counts on 3 from 58 and gets a chip)		
1	61 (after Player 2's +3)			Black 7 (counts on 7 from 61)	
1	68 (after Player 3's +7)				Queen (0) (stay at 68)
2	68 (after Player 4's 0)	Red 7 (counts back 7 from 68)			
2	61 (after Player 1's -7)		Red 5 •• (counts back 5 from 61 and gets another chip)		
2	56 (after Player 2's -5)			Black 3 (counts on 3 from 56)	
2	59 (after Player 3's +3)				Red jack •• (counts back 20 from 59 and gets two chips)
3	39 (after Player 4's -20)	Red 4 (counts back 4 from 39)			

ACTIVITY 4.9

Name: Number Catch   **Type:** Game

About the Game: *Number Catch* is a game for practicing counting on and back by different amounts. The *Number Catch* cards say *more* and *less* (e.g., 2 more or 7 less). Before students can play the game, they must understand that *more* means to count on from a number and that *less* means to count back. Look for students who associate the words with directions on the number chart, such as "*more* means to the right or up."

Materials: Fifteen game pieces (you can use centimeter cubes), a *Number Catch* board for two players, a set of 0–100 number cards, and two or three sets of *Number Catch* cards

Directions:

1. Players randomly place fifteen game pieces on the *Number Catch* board.

2. Players shuffle number cards and *Number Catch* cards and place them in two decks face down.

3. Players take turns flipping a *number card* and three *Number Catch* cards.

4. The players use the number card to count on or back by the amount shown on the *Number Catch* cards. Note that the cards can be used in any combination (see the example below).

5. If the result of the count matches a number on the chart that has a game piece, the player "catches" (i.e., removes) the game piece and their turn ends.

6. The game ends when all pieces are removed or time is up. The player with the most pieces wins the game.

> **TEACHING TAKEAWAY**
> Remember that resources used for all sorts of activities (such as 0–100 number cards or digit cards) are available online. If you can, print them on heavy bond paper and consider having volunteers cut and bag them for you.

For example, Player 1 flips the number card 27. They then flip the *Number Catch* cards 8 more, 1 less, and 30 more. There is no piece on 35 (8 more than 27), 26 (1 less than 27), or 57 (30 more than 27). However, there is a piece on 56. It can be found by counting on 30 from 27 (57) and then 1 less (56). Player 1 can remove the piece on 56. Then, it is Player 2's turn. Note that you can change the rules so that a player can remove multiple pieces with one turn and different combinations. In this example, Player 1 couldn't remove any other pieces, but if there was a piece on 65, they could remove it by combining 30 more and 8 more.

27

| 30 more | 1 less | 8 more |

Number Catch

Directions: Place fifteen pieces on the board. Make a two-digit number. Then, flip three Number Catch cards. Remove the pieces that match the results.

91	92	93	94	95	96	97	98	99	100
81	⬛	83	84	85	86	87	88	89	90
71	72	⬛	74	⬛	76	77	78	79	80
61	62	63	64	65	⬛	67	⬛	69	70
51	⬛	53	54	55	⬛	57	58	59	60
41	⬛	43	⬛	45	46	47	48	49	⬛
31	⬛	33	34	35	36	37	⬛	39	40
⬛	22	23	24	⬛	26	27	28	29	30
11	12	13	14	15	16	⬛	18	19	20
1	2	3	4	5	6	7	8	9	10

(Continued)

(Continued)

RESOURCE(S) FOR THIS ACTIVITY

Number Catch

Directions: Place fifteen pieces on the board. Make a two-digit number. Then, flip three Number Catch cards. Remove the pieces that match the results.

91	92	93	94	95	96	97	98	99	100
81	82	83	84	85	86	87	88	89	90
71	72	73	74	75	76	77	78	79	80
61	62	63	64	65	66	67	68	69	70
51	52	53	54	55	56	57	58	59	60
41	42	43	44	45	46	47	48	49	50
31	32	33	34	35	36	37	38	39	40
21	22	23	24	25	26	27	28	29	30
11	12	13	14	15	16	17	18	19	20
1	2	3	4	5	6	7	8	9	10

1 more	2 more	3 more	4 more
5 more	6 more	7 more	8 more
9 more	1 less	2 less	3 less
4 less	5 less	6 less	7 less
8 less	9 less	WILD	WILD

10 more	20 more	30 more	40 more
50 more	10 more	20 more	30 more
40 more	50 more	10 less	20 less
30 less	40 less	50 less	10 less
20 less	30 less	40 less	50 less

 This resource can be downloaded at **https://qrs.ly/psf6a5o**

ACTIVITY 4.10

Name: *300 Is Perfect* **Type:** *Game*

About the Game: This *300 Is Perfect* game blends estimating and counting. The goal of the game is simple: Players try to get as close to 300 as possible without going over. Each player starts at 0 and rolls a digit. That digit can represent a number of ones or tens. The player decides which it will be based on the number they are counting on from and the goal of getting close to 300 at the end of their eight turns. Along the way, they practice counting on by tens or ones from a variety of numbers. A good way to introduce the game is to set a goal of 100. But such a low target limits counting by tens. You can adjust the game to any start number and any target, such as starting at 200 and playing *500 Is Perfect*, or starting at 400 and playing *3,000 Is Perfect* (in this version, the digit can represent tens or hundreds). Some variations are shown in the game board examples.

Materials: A *300 Is Perfect* game board (optional), a 10-sided die or digit cards

Directions:
1. Players take turns rolling a digit.

2. The player decides whether the digit will represent the number of ones or tens.

3. The player begins at 0 and counts on the amount that they determined.

4. On the next turn, the player rolls again and decides how to use the number.

5. They count on that amount from the new number.

6. The player must roll exactly eight times.

7. The player closest to 300 (without going over) wins.

The image on the left shows an example of a round played. The player's final number is 268. The middle shows how you could modify it for a different starting point with a goal of 500. The right shows how you could practice counting by hundreds and tens with a goal of 3,000.

300 Is Perfect

Directions: Start at 0. Roll a number. Decide if that is the number of ones or tens to count on from 0. Roll again and count on from your new number. Get **as close to 300** as possible without going over.

Tens	Ones	Total
6		60
	9	69
7		139
6		199
	4	203
	7	210
5		260
	8	268

500 Is Perfect

Directions: Start at 200. Roll a number. Decide if that is the number of ones or tens to count on from 200. Roll again and count on from your new number. Get **as close to 500** as possible without going over.

Tens	Ones	Total

3,000 Is Perfect

Directions: Start at 0. Roll a number. Decide if that is the number of hundreds or tens to count on from 0. Roll again and count on from your new number. Get **as close to 3,000** as possible without going over.

Hundreds	Tens	Total

online resources This resource can be downloaded at **https://qrs.ly/psf6a5o**

ACTIVITY 4.11

Name: Shake and Spill   **Type:** Center

About the Center: Shake and Spill is a simple center with endless possibilities for counting and skip counting. To prepare for this center, you want to create a collection of Styrofoam cups (or something similar) with a range of numbers written on the bottom of each cup. For counting within 120, you might write 38, 99, 82, and so on. Counting within 500 might use 361 or 478.

You also need a collection of place-value disks for this center. On one side of the disk, you want to write a −, *B*, or another symbol to direct students to count back by the amount on that side of the disk. If you don't have place-value disks, you can write on two-colored counters or bingo chips. An online resource for printing and cutting is also available.

Materials: Assorted place-value disks, Styrofoam cups with numbers on them, recording sheet (optional)

Directions:
1. Students grab a handful of assorted place-value disks and put them in a cup.

2. Students then exchange cups.

3. Students flip over their new cup to reveal a starting number on the bottom of their cup while spilling out their place-value disks.

4. Students use the place-value disks to count from the number on the bottom of the cup. Note: When the disks are spilled, some will direct to count forward and some will direct to count back.

5. Optional: Students record their starting number, disks, and the final number.

The image shows a student who flipped over a cup with the number 41 written on it. In the cup, there were five disks with 10 on them and four disks with 1 on them. When the disks were spilled, four disks were to count on by 10, one to count back by 10, two to count on by 1, and two to count back by 1. The student recorded their disks and wrote the final number.

Shake and Spill is a good activity for practicing counting on and back by tens and ones. As students become proficient with these counts, it's time to move on to counting by groups of tens and groups of ones. You can modify games to do this by having students group like disks (e.g., 10, 10, 10, and 10) and recording a disk with that amount (e.g., 40) or you can write groups on two-colored counters (e.g., +40, −40).

Number I Started With	Disks to Count By	Number I Ended With
41	(10) (-10) (10) (1) (1) (-1) (10) (10) (-1)	71

online resources ⬏ A recording sheet can be downloaded at **https://qrs.ly/psf6a5o**

TEACHING TAKEAWAY

Centers such as Shake and Spill are the perfect tool for quality practice that students can do at home. You can send a few cups and some disks home and have students show family members how the activity works.

ACTIVITY 4.12

Name: Flip   **Type:** Center

About the Center: Flip is an activity that practices counting by groups. In it, students start with a random number and turn over Flip cards that charge them with counting on or back by groups of ones and tens.

Materials: A deck of Flip cards, a 10-sided die or playing cards (0–9 values only), Flip Recording Sheet and place-value disks (optional)

Directions:

1. Flip cards are shuffled and placed face down.

2. A student uses a die or playing cards to generate a starting number.

3. The player then turns over a Flip card, counts on or back as directed by the card, and determines the new number.

4. From the new number, the student turns over another Flip card and counts on or back as directed by the card to determine a new number.

5. The student tries to flip and count seven times without "busting" (arriving at 0 or less). Alternate: Student tries to flip as many cards as possible without busting.

6. Optional: Students write their counts on their Flip Recording Sheet as shown in the example.

The image shows a student's first round of Flip. Their starting number is 67. Then, they flip a count on (+) 20 card, landing at 87. On their next turn, they flip a count back (−) of 30, resulting in 57. The third flip gets them to 67 before their fourth flip causes them to bust.

Start	Flip 1	Now	Flip 2	Now	Flip 3	Now	Flip 4	Now	Flip 5	Now	Flip 6	Now	Flip 7	Now
67	+20	87	-30	57	+10	67	-80	Bust						

The player's second round of Flip works out better. They start with 37. They were then able to flip seven cards without busting (hitting 0).

Start	Flip 1	Now	Flip 2	Now	Flip 3	Now	Flip 4	Now	Flip 5	Now	Flip 6	Now	Flip 7	Now
37	+50	87	-40	47	+5	52	+10	62	-2	60	+80	140	-1	139

RESOURCE(S) FOR THIS ACTIVITY

Count On (+)	Count On (+)	Count On (+)
20	30	40
50	80	90
20	30	40
50	80	90

Count Back (−)	Count Back (−)	Count Back (−)
20	30	40
50	80	90
20	30	40
50	80	90

Count On (+)	Count On (+)	Count On (+)
1	1	1
2	2	2
5	5	5
10	10	10

online resources ➤ Flip cards, recording sheet, and place-value disks can be downloaded at **https://qrs.ly/psf6a5o**

Properties of Addition and the Inverse Relationship With Subtraction

PROPERTIES OF ADDITION OVERVIEW

The properties of the operations are perhaps the most important mathematical content to understand to enact reasoning strategies. Those critical properties for addition are listed in the table below.

PROPERTY	SYMBOLIC REPRESENTATION	HOW A STUDENT MIGHT EXPLAIN THE PROPERTY
Commutative	$a + b = b + a$	"You can add numbers in any order and get the same answer."
Associative	$(a + b) + c = a + (b + c)$	"When you add three numbers, you can add any pair first, then add the third number and you get the same total."
Inverse Relationship	If $a + b = c$ then $c - b = a$ & $c - a = b$	"Either number in an addition problem can be subtracted from the total and the answer is the other number."
Additive Identity	$a + 0 = 0 + a = a$	"Adding zero to any number means you get the same number you started with."

Source: Adapted from Van de Walle et al., 2023.

To be used for reasoning strategies, mathematical properties and relationships must be deeply understood. For many students, properties are distilled into statements such as $a + b = b + a$ or "addition is the opposite of subtraction," and it ends there. The outcome of this shallow understanding is a student who might be able to answer 9 + 3 automatically but struggles with 3 + 9 or 12 – 9. Being able to recognize and name a property is one thing, but properties lose their value if they are simply rote, disconnected recitations or gross generalizations without a useful purpose. What makes properties valuable is when students can employ them automatically.

This is not to say that students shouldn't be exposed to symbolic representations of properties (e.g., $a + b + c = a + c + b$) or that they shouldn't be asked to describe properties and the inverse relationship in their own words—they should! But students must be able to apply properties to compute efficiently and flexibly.

> **TEACHING TAKEAWAY**
>
> Properties must be well-understood and automatic for students to be able to compute efficiently and flexibly.

HOW DOES THIS MODULE CONTRIBUTE TO FLUENCY STRATEGIES?

The properties of addition—and addition's inverse relationship to subtraction—essentially make the significant strategies (Bay-Williams & SanGiovanni, 2021) for addition and subtraction possible. We use properties to manipulate, think flexibly, and find efficiencies. The commutative property comes alive when we reverse the order of addends to count on more efficiently. Consider 58 + 76. A student can count on from the first addend (as shown on the left) but another student can reverse the addends, counting on from the larger addend. While either is appropriate, the latter can make for easier counting.

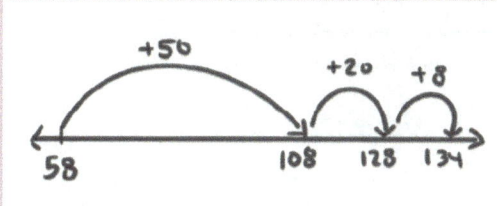
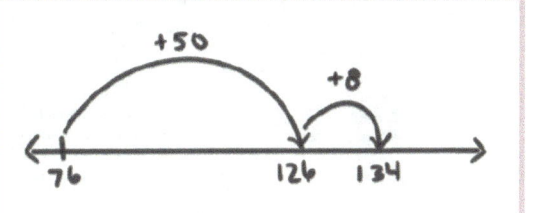

The associative property can be applied to the Make Tens strategy (shown on the left below) when a student decomposes an addend and then rearranges the parts. In this problem, 47 + 35 becomes 47 + 32 + 3. Then, the property is used to show how to make a ten. The inverse relationship between addition and subtraction is often called on and used in the Think Addition strategy. The work on the right shows Think Addition for 291 – 269. The problem becomes 269 +? = 291. The student counts up 20 and then 2 to find a difference of 22.

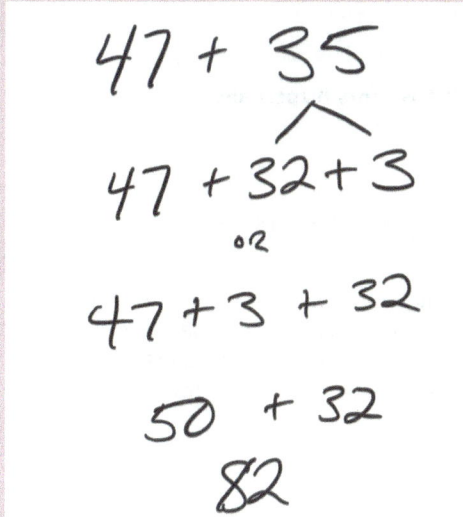
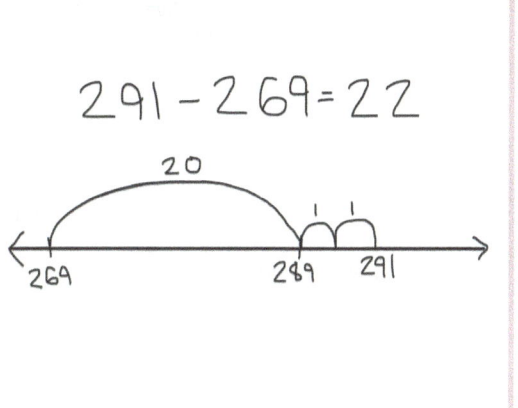

The properties of addition reference page provides an overview, important representations, connections to fluency, and actions to avoid. It can be downloaded and used for reference in planning, teaching, and discussions with colleagues and families.

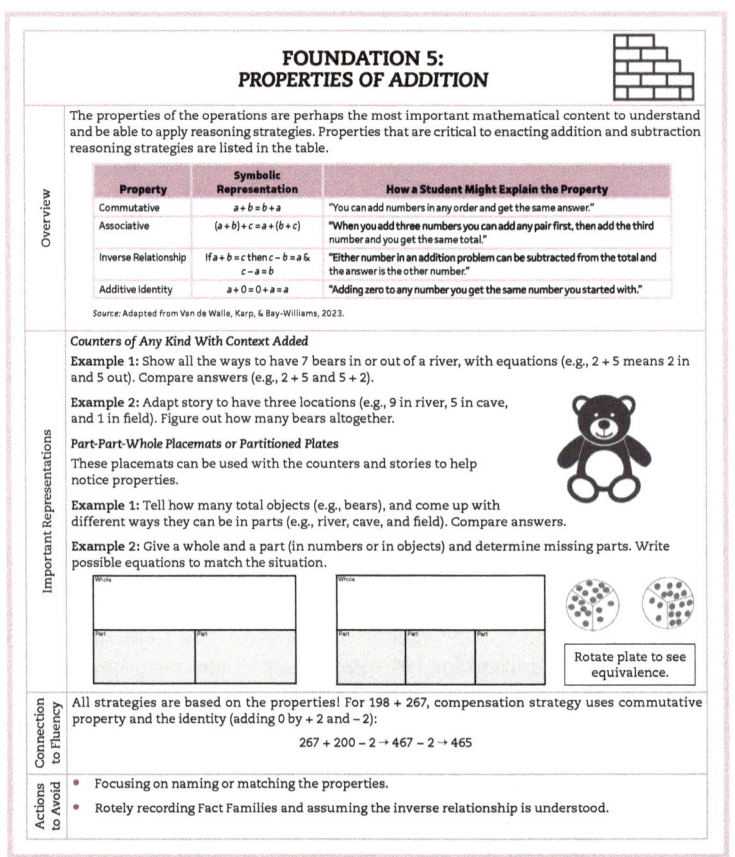

The following is the content of the Foundation 5 reference card shown above:

FOUNDATION 5: PROPERTIES OF ADDITION

Overview

The properties of the operations are perhaps the most important mathematical content to understand and be able to apply reasoning strategies. Properties that are critical to enacting addition and subtraction reasoning strategies are listed in the table.

Property	Symbolic Representation	How a Student Might Explain the Property
Commutative	$a + b = b + a$	"You can add numbers in any order and get the same answer."
Associative	$(a + b) + c = a + (b + c)$	"When you add three numbers you can add any pair first, then add the third number and you get the same total."
Inverse Relationship	If $a + b = c$ then $c - b = a$ & $c - a = b$	"Either number in an addition problem can be subtracted from the total and the answer is the other number."
Additive Identity	$a + 0 = 0 + a = a$	"Adding zero to any number you get the same number you started with."

Source: Adapted from Van de Walle, Karp, & Bay-Williams, 2023.

Important Representations

Counters of Any Kind With Context Added

Example 1: Show all the ways to have 7 bears in or out of a river, with equations (e.g., 2 + 5 means 2 in and 5 out). Compare answers (e.g., 2 + 5 and 5 + 2).

Example 2: Adapt story to have three locations (e.g., 9 in river, 5 in cave, and 1 in field). Figure out how many bears altogether.

Part-Part-Whole Placemats or Partitioned Plates

These placemats can be used with the counters and stories to help notice properties.

Example 1: Tell how many total objects (e.g., bears), and come up with different ways they can be in parts (e.g., river, cave, and field). Compare answers.

Example 2: Give a whole and a part (in numbers or in objects) and determine missing parts. Write possible equations to match the situation.

Rotate plate to see equivalence.

Connection to Fluency

All strategies are based on the properties! For 198 + 267, compensation strategy uses commutative property and the identity (adding 0 by + 2 and – 2):

$$267 + 200 - 2 \rightarrow 467 - 2 \rightarrow 465$$

Actions to Avoid

- Focusing on naming or matching the properties.
- Rotely recording Fact Families and assuming the inverse relationship is understood.

ASSESSING PROPERTIES OF ADDITION

Assessing the properties is *not* about naming them or completing a matching activity where the property names are matched to an example. The key to assessing is to see if students are appropriately *applying* the properties. When you ask students if an addition or subtraction action always works (and ask them to justify their answer), you are likely going to gain strong insights on their understanding of the properties for addition.

ADDITION PROPERTY LOOK-FORS

With properties, there is a lot to look for. For the commutative property, students might match two equations, such as 4 + 9 and 9 + 4, because they recognize that in both cases, they are combining the same two numbers for a whole. With three or more addends, there is more to look for, as students can move numbers to make tens or in other ways to make the sum easier to compute. In a similar way, rather than citing rote fact families, you are looking for an understanding of the inverse relationship between addition and subtraction, recognizing the role of parts and the whole. You should find that your students can

- explain how and why properties work,
- identify and create examples of properties,
- rearrange or group addends to make addition more efficient through the commutative and/or associative property,
- recognize and be able to explain why commutative and associative properties work for addition, but not for subtraction; and
- solve problems in different ways (i.e., solve a part-unknown story problem as a missing addend problem and a subtraction problem).

The following prompts provide insights into whether students intuitively understand the properties (i.e., they know that the order in which you add numbers does not matter) and the extent they are able to apply the properties. These prompts can be traded out to explore other properties. More prompts are provided in Activity 5.5.

Pose an example of the commutative property (e.g., 13 + 18 = 31 and 18 + 13 = 31). Ask the student to explain how the equations are related.	Tell the student that 26 + 38 = 64. Ask them what 38 + 26 is (consider writing this down or allowing them to write it down). Ask them how they found 64. If they cite the commutative property, have them create a new example.	Tell the student that 46 + 45 = 91. Ask them how that could help them solve 91 – 45. Listen for them to describe the relationship. Then ask students to solve a subtraction problem (e.g., 64 – 49 or 224 – 198) and listen for them using the related inverse equation.
Write some equations on index cards (e.g., 18 + 16 + 14, 10 + 44 + 6 + 17, 27 + 37 + 13). Ask the student to find the sum on the first card. If they add left to right, ask if they can add it another way. Have students use parentheses to show how they grouped numbers. Repeat for second and third card.	Write down 9 + 47 + 1. Ask the student to find the sum. Look to see if they rearrange the addends. If they do, ask them how they know they can do that. Then, repeat with a new problem and less obviously compatible numbers (e.g., 16 + 14 instead of 9 + 1).	Provide addition and subtraction equations. Have the student match the equations and tell how they're related.

EXPLICIT INSTRUCTION FOR PROPERTIES OF ADDITION AND THE INVERSE RELATIONSHIP WITH SUBTRACTION

Students need to explore properties and make their own arguments as to why the properties work. It can be tempting to simply tell students that you can flip-flop addends, but for them to apply a property, they must have their own understanding. It's fine if they say that the addends flip-flop. When they do, confirm that the order of addends can be reversed without changing the sum. Challenge them to tell stories and create examples to show their understanding. For example, a student might explain that they have four crayons in one hand, and three crayons in the other hand—if they cross their hands, it is still the same amount!

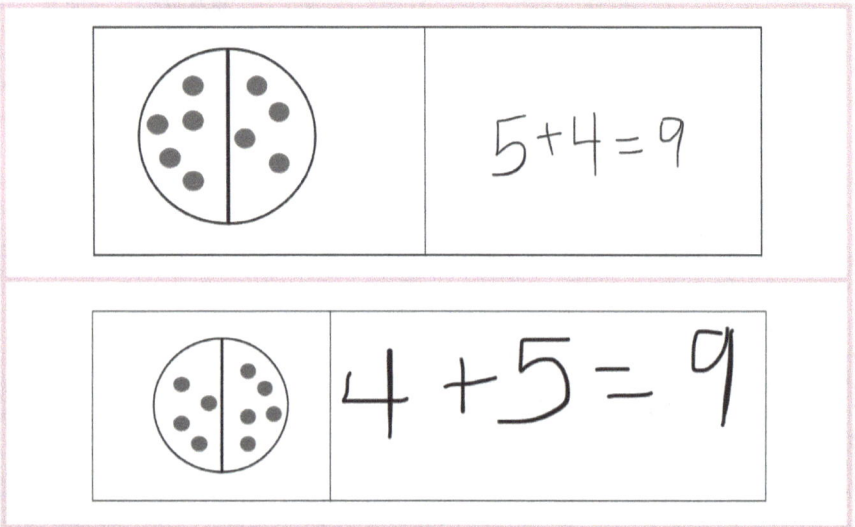

ACTIVITY 5.1
TWIST IT!

This activity will provide opportunities for students to experience adding numbers in any order. They will rotate through stations, working in pairs or small groups. Students who do not yet fully understand the commutative property need multiple exposures to situations to help them start to comprehend the idea that we get the same answer no matter in which order we add.

Prepare paper plates by dividing them into two equal parts, drawing a line down the middle of each plate or using colored masking tape (painter's tape). Place one plate, along with a predetermined number of counters, at multiple stations around the room. Students will need paper and a pencil to record their equations.

At each station, one partner will put all of the counters in their hand and drop them over the plate. (If a counter happens to fall exactly on the line, the students should move it to one side or the other.) Then, each student writes the equation that matches the number of counters shown on each side of the plate. They write the equation on their own and then compare their work with their partner. Once they agree on the equation, they say, "Twist it" and work together to rotate the plate so it ends up being the opposite of what it was originally. The students will write the new equation directly underneath their first equation. Once both partners have the two equations written and agree they are correct, they clear the plate of the counters and move on to a new station.

After the students move through a number of rotations, bring the class together to discuss what they noticed about the equations they wrote. Make sure to reinforce the idea that numbers can be added in any order. Students may use informal language such as "the numbers switched around" or "they flipped around," which is fine, as it helps them better understand what is happening.

TEACHING TAKEAWAY

This activity is described with single-digit addends and counters. Later experiences can make use of two-digit addends and changing the representations from counters to sketches of sticks and dots showing base-ten blocks.

This activity can and should be extended to equations with three addends (this is when the commutative property becomes really useful!). Rather than dividing a paper plate into two parts, draw lines to divide each plate into thirds (or fourths); paper plates that are already divided into three sections could also be used. The students will drop the counters over the plate and record the equation. Tell students to find three expressions, twisting each plate to support their reasoning.

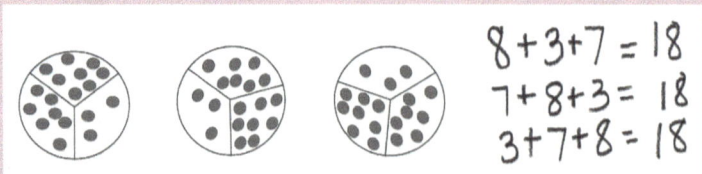

Ask students, "Which of these expressions is in the order you would like to add the numbers?" Repeat this activity, but have students twist the plate to select what they think is the best order for the addends and then solve it.

ACTIVITY 5.2
MINGLE AND MATCH

This instructional activity helps students recognize problems that are mostly similar, with the only difference coming from the application of the commutative property. For example, students will notice that 4 + 5 and 5 + 4 are the same problem, but the addends are in a different order or "switched." During Mingle and Match, students will get to look at different problems and determine if they are the same or not.

To begin, prepare a set of cards with expressions with two addends. Use index cards (or something similar) to make sets of cards with problems similar to those in the table below. Each card will have a corresponding card but the addends will be in a different order. An example of a pair would be 6 + 3 and 3 + 6.

Distribute one card to each student. Have students position themselves around the room, as they will be moving during this activity. Play music and have students walk around the room (safely!) and exchange cards with students as they walk by. They will get cards from each other until the music stops. When it stops, they will find the person that is holding the card that matches their card. Once they make a match, the pair sits down. When all the students are sitting down, the teacher begins playing the music again and all students return to walking slowly around the room and exchanging cards with one another. Repeat this activity with expressions that have three or four addends. Once partners find each other, ask them to decide which order they prefer for solving the problem and solve it that way (comparing their answer with their partner's answer).

Another adaptation of this activity would be to make cards that include related addition and subtraction facts. A pair of cards for this variation might be 4 + 5 = 9 and 9 − 5 = 4. This activity could eventually become a center. Using one set of cards, students could find the cards that match.

EXAMPLES WITH TWO ADDENDS		EXAMPLES WITH THREE ADDENDS		EXAMPLES FOR RELATED ADDITION AND SUBTRACTION FACTS	
Card	Matches With	Card	Matches With	Card	Matches With
6 + 4 =	4 + 6 =	7 + 5 + 3 =	7 + 3 + 5=	3 + 7 =	10 − 7 =
8 + 3 =	3 + 8 =	2 + 4 + 8 =	2 + 8 + 4 =	6 + 8 =	14 − 8 =
7 + 4	4 + 7	9 + 7 + 1 =	9 + 1 + 7 =	7 + 5 =	12 − 7 =

While these are single-digit expressions, this activity can grow to two- and three-digit sums. To highlight the associative property, you can vary this activity by creating cards with three or four addends, grouping different sums together: 19 + (11 + 14) and (19 + 11) + 14 or, with four addends (39 + 12) + (8 + 11) and 39 + (12 + 8) + 11.

ACTIVITY 5.3
WHICH SWITCH?

Materials: Which Switch? cards

When given three or more addends, students often default to adding the numbers in the order they are presented. In some cases, this is the most logical approach, but sometimes it is not. This activity will help students identify when they want to add in the given order and when they do not. Students might look for two addends that make a decade number or other combinations they know. It may be that addends need to switch places to make this happen. Thus, students explore both the commutative and associative properties through this task.

For each group of two or three students, download or create cards with problems containing three addends. You can use the suggestions given or create your own. (The expressions in the left two columns would be better as students initially explore the idea of associativity. Once students become more comfortable with adding larger numbers, you could move to the number combinations on the right.)

SUMS WITHIN 30		SUMS GREATER THAN 30	
14 + 9 + 6	14 + 6 + 7	73 + 18 + 17	27 + 23 + 32
8 + 2 + 9	9 + 11 + 5	56 + 24 + 13	48 + 15 + 22
16 + 8 + 4	12 + 8 + 3	77 + 28 + 33	27 + 13 + 50
3 + 11 + 7	5 + 2 + 15	19 + 41 + 47	61 + 19 + 17
2 + 9 + 18	19 + 6 + 1	55 + 19 + 35	52 + 35 + 18

Introduce the task to students by saying, "Sometimes it helps to switch the order that you add." The students will work together to sort the cards into two piles: problems that might benefit from a switch to solve and those that would not. There are two ways to switch. First, they can move the numbers around: for example, they may say that 14 + 9 + 6 would benefit from a switch, as 14 + 6 + 9. Second, they can group to make a switch (i.e., to add the last two numbers first). For example, for 5 + 20 + 15, a student might want to add the larger numbers first and thus will switch the order by grouping: 5 + (20 + 15). As students work, probe their thinking by asking questions of why they placed a particular problem into the *switch* or *don't switch* group.

Once the students have sorted their cards into the two piles, bring the group together for a discussion. It might be best for you to have a set of cards to use during the discussion. Have the students share their discoveries about what they noticed about the different problems. Make sure to bring attention to those combinations where switching made the computation easier (as it allowed students to work with benchmark numbers).

RESOURCE(S) FOR THIS ACTIVITY

14 + 9 + 6	8 + 2 + 9	16 + 8 + 4	73 + 18 + 17	56 + 24 + 13	77 + 28 + 33
3 + 11 + 7	2 + 9 + 8	14 + 6 + 7	19 + 41 + 47	55 + 19 + 35	27 + 23 + 32
9 + 11 + 5	12 + 8 + 3	15 + 20 + 15	48 + 15 + 22	27 + 13 + 50	61 + 19 + 17
19 + 6 + 1	7 + 13 + 7	17 + 7 + 3	52 + 35 + 18	37 + 13 + 27	46 + 23 + 24

 This resource can be downloaded at **https://qrs.ly/psf6a5o**

ACTIVITY 5.4
THEY'RE ALL RELATED!

It is important for students to understand the inverse relationship between addition and subtraction. Once students become comfortable with this, they can use related facts to solve problems. It isn't uncommon for students to use addition to help them solve subtraction problems. A student might look at 18 – 9 and think, "I know 9 + 9 is 18, so 18 – 9 is 9."

During this activity, students will be organized into groups of four. The group will sit in a circle and will have each person write on something, such as a whiteboard or piece of paper. They will draw a box divided into four equal parts. Each group of students will need a set of playing cards with the tens and face cards removed. The cards are placed face down in the middle of the group. The first leader chooses four cards and lays them out in the middle of the group while saying the two numbers that are built (by combining two cards) out loud. Once the numbers are announced, each person in the group writes either an addition or subtraction equation using the two numbers as parts. Once they have an equation written, they put their thumbs up and rest their hands on their knees. The leader tells the group to switch once they see that everyone is giving a thumbs up. The group hands their board to the person on their left. They read the equation on the board and add a different but related equation. This process continues for four rotations, at which point each board should have four related equations written on it. A board might look like this at the end of a round when the numbers 29 and 44 are used:

| 73 – 29 = 44 | 44 + 29 = 73 |
| 29 + 44 = 73 | 73 – 44 = 29 |

Once the round is completed, all group members turn their work toward the middle of the circle so other group members can see the work on all four boards. Then, the leader gives the group a compliment for working together: "We all worked together to find the related facts for 29 and 44. Good effort!" The person to the left of the original leader becomes the new leader. The group repeats the process using two new numbers each time.

This activity can be simplified when initially introducing the game. The tens could be left in the deck. Students would only turn over two cards and use the value of those cards as addends. This would give students practice with number combinations up to twenty. Another adaptation of the activity would be to provide a directive that the first equation written on a board is addition. The rule could be implemented that the next person adds the other related addition equation. This would mean the last two group members write subtraction equations.

ACTIVITY 5.5
PROMPTS FOR PROPERTIES OF ADDITION AND THE INVERSE RELATIONSHIP WITH SUBTRACTION

Use the prompts below as opportunities to develop understanding of and reasoning with the strategy. Have students use representations and tools to justify their thinking, including base-ten models, number lines, number charts, and so on. After students work with the prompt(s), bring the class together to exchange ideas. Remember that these could be useful for collecting evidence of student understanding.

1. The problem is 37 + 28 + 13. Tara says she can add 37 + 13 first to make 50 and then add 28 to make 78. Does Tara's strategy work? Do you think it's a good idea?

2. Kallah said 47 + 23 isn't the same as 23 + 47. How might you convince her the sum will be the same?

3. Chris built a tower by stacking eight blue blocks, then nine yellow blocks, and then eleven red blocks. Would the tower be the same height if Chris started with the nine yellow blocks, then eleven red blocks, and then eight blue blocks? How do you know?

4. Mia uses the commutative property to think of 75 + 32 as 57 + 23. What would you tell her about her thinking?

5. Dex says that you can solve any subtraction problem by thinking about addition and adding up. Create examples that prove whether you agree or disagree with Dex.

6. The chart shows related equations. Make a chart with five new sets of related equations. In each problem, two of your numbers have to be greater than 10.

29 + 17 = 46	17 + 29 = 46	46 – 17 = 29	46 – 29 = 17

7. Using three or four addends, create an example of the commutative property and an example of the associative property. Explain how the two properties are different.

8. Create a problem that would be good for solving by applying the associative property. Explain why you think your problem is a good choice.

QUALITY PRACTICE FOR PROPERTIES OF ADDITION AND THE INVERSE RELATIONSHIP WITH SUBTRACTION

Quality practice has to be engaging to be effective (Bay-Williams & SanGiovanni, 2021). Obviously, you want practice to be fun, but remember to keep it brief. A short burst of enjoyable practice done over a long period of time will have a much better effect than having students practice a skill once a week for forty minutes at a time. This is especially important for the fluency foundations in this book. That's why we have included collections of routines, games, and centers throughout this book. Each of these practice activities is intended to engage students in short, 10- to 15-minute bouts of practice carried out over a few weeks or months.

ACTIVITY 5.6

Name: "That's the Truth"　　　　　　　　**Type:** Routine

About the Routine: That's the Truth is a routine for students to practice properties of addition and the inverse relationship between addition and subtraction. Partners are presented with a set of statements and they must decide which statements are true. But students should not find sums and differences to determine which are true. Instead, they have to ground their arguments in their understanding of properties, relationships, and number composition. After sharing some of their proofs, the teacher poses a "so then" statement to apply their proven-true statement in another problem. For the upper-left example, the true statement is C: 43 + 19 is the same as 19 + 43. A "so then" statement after discussion might be, "So then, 27 + 65 is the same as ___." Note that you can choose to have more than one statement in the routine be true, as shown in the right-hand examples.

Materials: This routine does not require any materials.

Directions:　1. Pose statements to students for them to consider, as shown in the examples.

　　　　　　　2. Ask students, "Which of these statements are true?"

　　　　　　　3. Have partners discuss which they think is true and why.

　　　　　　　4. Bring the class together to talk about their ideas.

　　　　　　　5. Come to agreement as a class about which statement is true.

　　　　　　　6. After it is clear which statement(s) is true, have students create a new statement with their partner that is similar to the true statement.

> **TEACHING TAKEAWAY**
> Labeling expressions and equations can help you facilitate conversations with students about them.

Templates are available for download, though it's just as easy to write a few statements on your board for students to discuss. The images show some ideas focused on the commutative property. This routine is also effective with the associative property or mixing in different properties.

(Continued)

(Continued)

The upper-left example features the commutative property. The two false statements play on reverse digits and an operation. The upper-right example shows how the associative property could be examined. The lower-left example is worded differently for framing the inverse relationship between addition and subtraction. The lower-right example shares a different approach so students can grapple with decomposing addends. Also note that there can be more than one true statement.

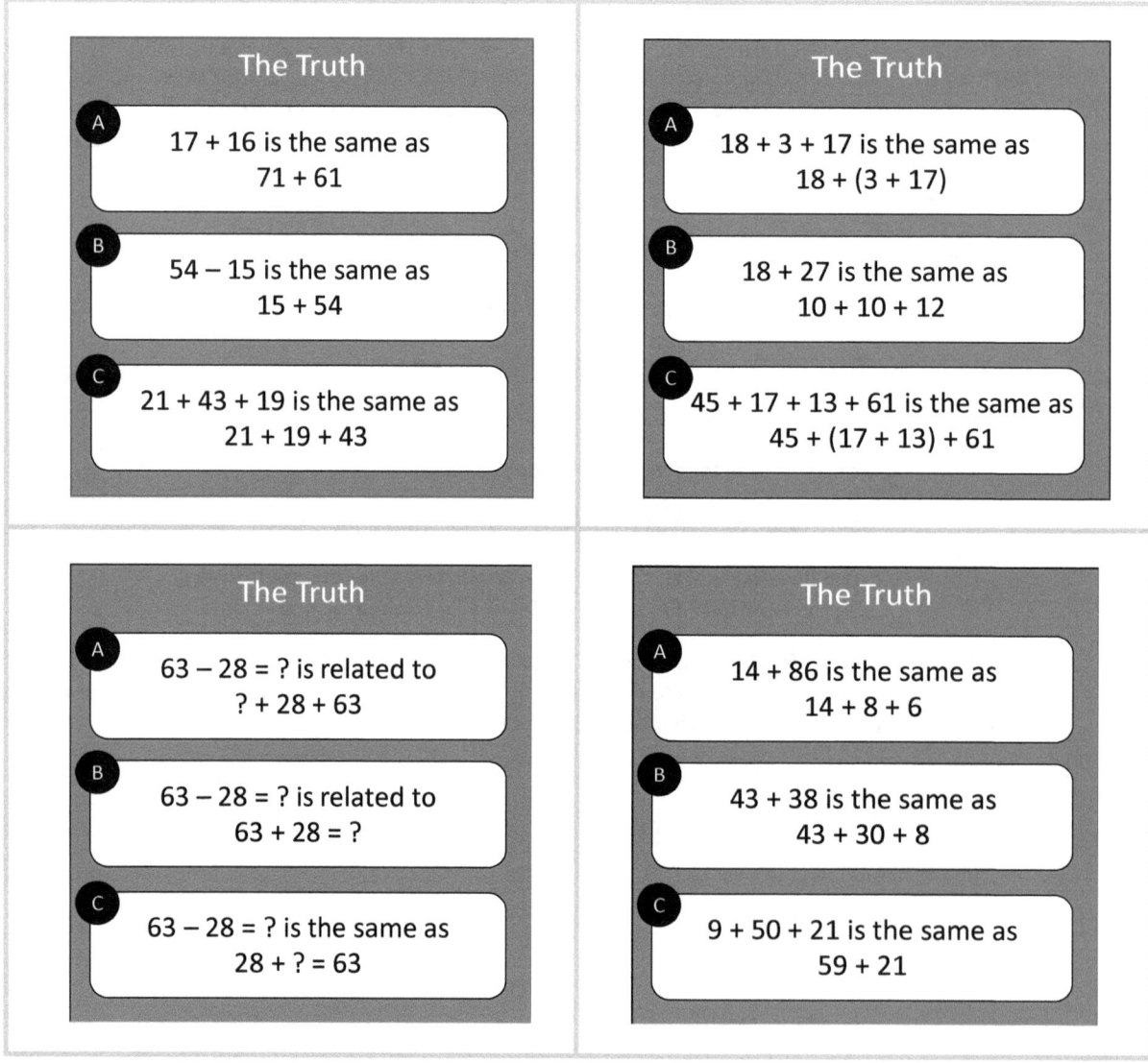

The prompts for this activity can be downloaded at **https://qrs.ly/psf6a5o**

Remember that you want students to describe why a statement is true based on their reasoning and understanding of the property. So, in the upper-left example, you want students to say that the order of the addends doesn't matter (or something similar). You don't want them to say, "C is true because 62 is the same as 62."

ACTIVITY 5.7

Name: The Match   **Type:** Game

About the Game: The commutative and associative properties of addition play an important role when students are learning to add. Students need to become comfortable recognizing and applying these properties when they add numbers together. This game will help students recognize problems that contain the same numbers, but in a different order. Students will play a game similar to Memory, where they have to find pairs of cards that match. Matching pairs will have expressions that contain the same numbers, but the order is switched.

Materials: *The Match* cards

Directions:

1. Two students lay a set of 24 cards face down in a 6 × 4 array.

2. Player 1 turns over two cards. If the cards do not have expressions that match, the student turns the cards back over to their original spot in the array and the player loses their turn. If the cards do have matching expressions, the player says, "The Match!" and keeps the two cards.

3. Players take turns until all cards are removed from the array. The player with the most cards at the end of the game is the winner.

4. Cards can be combined, shuffled, and put back in a new 6 × 4 array to play again.

5. A variation of the game would be that once a player finds a match, they have to say the correct sum to keep the cards. Once the correct answer is given, that player gets another turn. If a player says an incorrect answer, the cards are turned back over and the player loses their turn.

> **TEACHING TAKEAWAY**
>
> You may find that students evaluate the expression to compare and determine matches rather than considering the property represented. To avoid this, provide the sum so that students are directed to examine the addends and consider the properties rather than relying on calculating.

For example, a player selects cards 4 + 5 and 3 + 7; since they are not a match, the cards are turned face down again and it is the next player's turn. If a 4 + 3 + 7 and 3 + 4 + 7 is revealed, the player gets to keep the cards. When playing the variation of the game, if the player says "14" as the answer to the equation, that player can keep the cards and turn over two new cards.

RESOURCE(S) FOR THIS ACTIVITY

The Match (0–10)				The Match (10–20)		
5 + 4	4 + 5	3 + 6		5 + 3	6 + 5	6 + 7
6 + 3	7 + 2	2 + 7		7 + 6	7 + 8	8 + 7
8 + 2	2 + 8	7 + 3		8 + 3	3 + 8	8 + 4
3 + 7	6 + 4	4 + 6		4 + 8	8 + 5	5 + 8

 Card sets showing number combinations up to 10, number combinations up to 20, and number combinations with three addends can be downloaded at **https://qrs.ly/psf6a5o**

ACTIVITY 5.8

Name: *Three Numbers for 10 Points*　　　　　　　　　**Type:** *Game*

About the Game: This game mixes the associative property with a touch of chance. In it, players add three numbers (using grouping symbols to show how they added, if appropriate) and compare their sum to an amount they spin, with a goal of gaining points based on the comparison. You can choose to have students record their equations on a piece of paper, whiteboard, or downloadable recording sheet. After students play the game, be sure to talk about not only why the property works but when or what problems were good for using it. For example, a student who has 7 + 8 + 2 would find it useful to group the numbers, creating 7 + (8 + 2) to make a ten.

Materials: *Three Numbers for 10 Points* spinner game board, three 10-sided dice or playing cards (face cards removed and aces = 1), and *Three Numbers for 10 Points* recording sheet (optional)

Directions:　1. Players take turns generating three numbers by rolling a die three times or drawing three cards.

　　　　　　　2. Players record their numbers in the order they rolled/drew them and write an expression.

　　　　　　　3. Players determine the order they will add their numbers and use parentheses (as needed) to illustrate when they are not adding from left to right.

　　　　　　　4. Players find the sum of their three numbers.

　　　　　　　5. Players spin the spinner to determine their score for that round:

- one point if their sum is less than the number they spin

- two points if their number is equal to the number they spin

　　　　　　　6. The first player to score 10 points wins.

The game board on the left is used with numbers 1 through 10. A student might create 4 + 10 + 6 and group 10 + 6 to add the larger values first. If they spin 30, they get 1 point because their sum is less than 30. If they spin 20, they get 2 points because their sum is equal to 20. However, if they spin 10, they don't get any points because 20 is greater than 10. The left side shows a game where players are keeping score in the space provided with tally marks. Their equations can be written on the downloadable recording sheet. The right side shows how the game can be modified for adding larger two-digit numbers. Number cards for that version are available online with the game boards and the optional recording sheet.

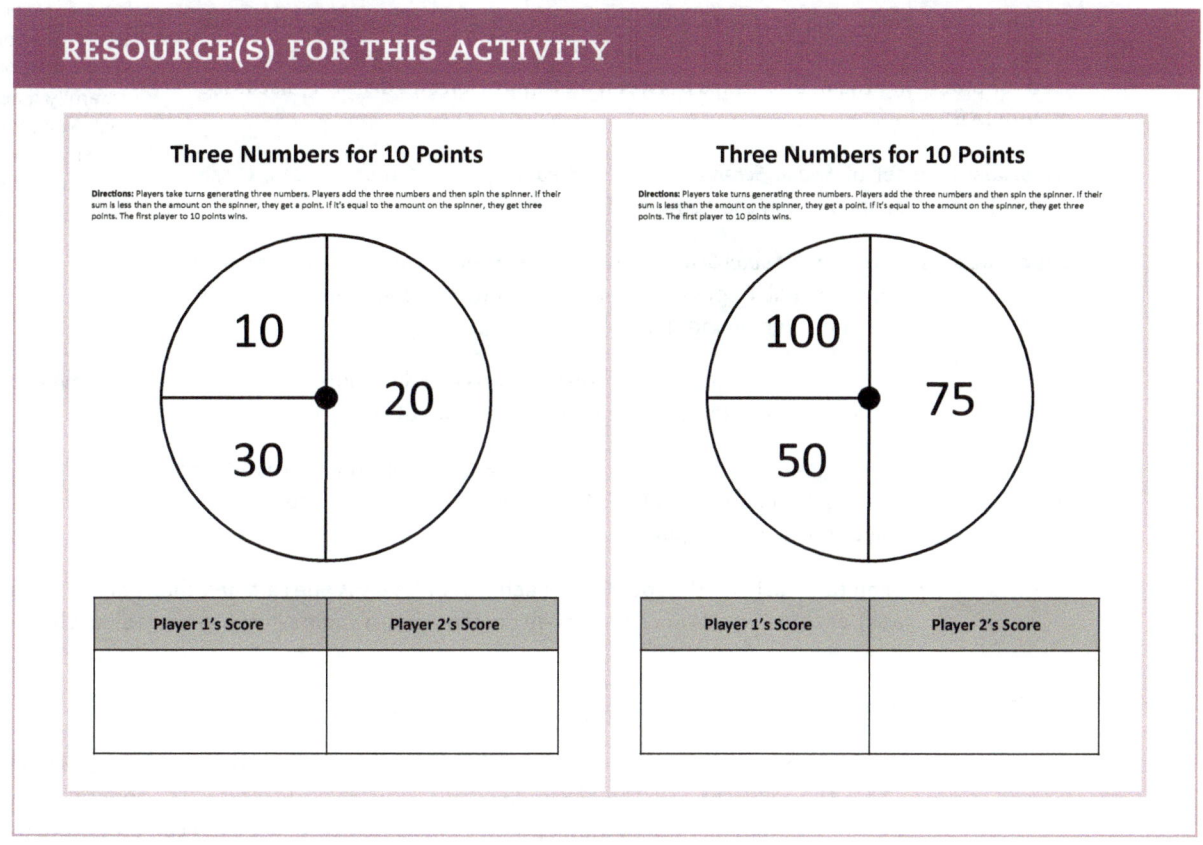

RESOURCE(S) FOR THIS ACTIVITY

Three Numbers for 10 Points

Directions: Players take turns generating three numbers. Players add the three numbers and then spin the spinner. If their sum is less than the amount on the spinner, they get a point. If it's equal to the amount on the spinner, they get three points. The first player to 10 points wins.

10
20
30

Player 1's Score	Player 2's Score

Three Numbers for 10 Points

Directions: Players take turns generating three numbers. Players add the three numbers and then spin the spinner. If their sum is less than the amount on the spinner, they get a point. If it's equal to the amount on the spinner, they get three points. The first player to 10 points wins.

100
75
50

Player 1's Score	Player 2's Score

online resources

This resource can be downloaded at **https://qrs.ly/psf6a5o**

ACTIVITY 5.9

Name: *Spoon Scramble*　　　　　　　　　　　　　**Type:** *Game*

About the Game: *Spoon Scramble* helps reinforce the relationship between addition and subtraction. It is easily adapted from basic fact equations to problems with multi-digit whole numbers. The goal is to collect three related equations and then grab a spoon! Then, players solve the equations and play again. This game can also be used for practicing the relationships between multiplication and division (discussed in Module 6).

Materials: One set of *Spoon Scramble* cards for each group of four players, three spoons (one fewer spoon than the number of players)

Directions:

1. The goal of *Spoon Scramble* is to make a set of three related equations and then grab a spoon (or notice someone else has grabbed a spoon and you grab a remaining spoon).

2. In a group of four players, mix up the cards and distribute three cards to each player so there are no remaining cards. Place three spoons in the middle of the table.

3. To prepare to play, players look to see if any of their three cards are related equations with the same missing value. If yes, that might be what they are going to collect (similar to collecting sevens in the classic game of Spoons).

4. Simultaneously, on the count of 3, each student passes one card and then checks to see if they have a set of three equation cards. If they do, they grab a spoon. If not, they decide which card they will pass next.

5. Repeat Step 4 until someone gets a full set and grabs a spoon.

6. Each person with a spoon finds the solution for the question mark (?) in the set of played equations. Check the solutions with the other players or a calculator. Players with spoons who correctly solve for the question mark score three points. The person with no spoon can only score one point with a correct answer.

7. Collect the cards, mix them up, and play again.

8. The first player to earn 16 points wins.

> **TEACHING TAKEAWAY**
>
> Simple, familiar games such as *Spoon Scramble* are perfect to introduce at family math nights so that students can play with friends, family, and caregivers for extra practice.

The images show four of the card sets available online. You can easily create new sets by editing the online versions or by giving each student three index cards and having them write three related equations similar to the examples. The upper-left shows basic fact cards, the upper-right shows sums within 30, the lower-left shows sums within 100, and the lower-right shows sums within 200. There is no limitation to the problems you use. The downloads shown can be edited for any type of number.

RESOURCE(S) FOR THIS ACTIVITY

SPOON SCRAMBLE SET 1

2 + ? = 11	11 − ? = 2	11 − 2 = ?
7 + ? = 15	15 − ? = 7	15 − 7 = ?
8 + ? = 14	14 − ? = 8	14 − 8 = ?
9 + ? = 16	16 − ? = 9	16 − 9 = ?

SPOON SCRAMBLE SET 2

12 + ? = 21	21 − ? = 12	21 − 12 = ?
17 + ? = 25	25 − ? = 17	25 − 17 = ?
18 + ? = 24	24 − ? = 18	24 − 18 = ?
19 + ? = 26	26 − ? = 19	26 − 19 = ?

SPOON SCRAMBLE SET 3

49 + ? = 76	76 − ? = 49	76 − 49 = ?
33 + ? = 51	51 − ? = 33	51 − 33 = ?
52 + ? = 81	81 − ? = 52	81 − 52 = ?
68 + ? = 85	85 − ? = 68	85 − 68 = ?

SPOON SCRAMBLE SET 4

84 + ? = 121	121 − ? = 84	121 − 84 = ?
66 + ? = 162	162 − ? = 66	162 − 66 = ?
83 + ? = 160	160 − ? = 83	160 − 83 = ?
75 + ? = 123	123 − ? = 75	123 − 75 = ?

online resources This resource can be downloaded at **https://qrs.ly/psf6a5o**

ACTIVITY 5.10

Name: Flip and Fill   **Type:** Game

About the Game: *Flip and Fill* is an easy game students can play to practice with the commutative property of addition or the inverse relationship between addition and subtraction. The goal of the game is to complete all of the statements on your game board before your opponent does. You'll also notice that this game has a nice question at the bottom of the game board for students to process what they did in the game. Their play, along with the explanations they offer after the game, can give you good insight into their understanding of the properties and relationships. You can even extend the question by having students create new examples to illustrate their thinking. *Flip and Fill* can be turned into an independent game or center by having students try to fill the board with the fewest number of cards.

Materials: Two sets of digit cards (0–9) or one deck of playing cards (tens and face cards removed, aces = 1) per player, *Flip and Fill* game board

Directions: 1. Players shuffle their cards and place them in a deck face down.

2. Players take turns flipping a card and finding a place on the game board where that number fits.

3. If a number can't be played, the player loses their turn.

4. The first player to fill all of the spaces on their game board wins.

These four examples of the game are available online. The two top examples show how it can be played with the commutative property using single-digit numbers (left side) or multi-digit numbers (right side). The bottom examples show how it can be used for the commutative property or for the inverse relationship between these operations.

Flip and Fill

Directions: Players take turns flipping a card and using it to fill a missing number.

9 + 3 is the same as 3 + ____	5 + ___ is the same as 6 + 5
___ + 8 is the same as 8 + 4	7 + ___ is the same as 6 + 7
5 + ___ is the same as 3 + 5	3 + 7 is the same as 7 + ___
6 + 8 is the same as 8 + ___	6 + 7 is the same as ___ + 6
7 + ___ is the same as 9 + 7	5 + 8 is the same as 8 + ___
___ + 9 is the same as 9 + 8	8 + ___ is the same as 2 + 8
9 + 4 is the same as ___ + 9	5 + 4 is the same as 4 + ___

On the back, explain how you knew how to fill in the missing numbers in each statement.

Flip and Fill

Directions: Players take turns flipping a card and using it to fill a missing number.

91 + 13 is the same as 13 + 9__	58 + 1__ is the same as 16 + 58
3__ + 38 is the same as 38 + 34	72 + 3__ is the same as 36 + 72
45 + 3__ is the same as 35 + 45	32 + 77 is the same as 77 + 3__
64 + 18 is the same as 18 + __4	26 + 57 is the same as 5__ + 26
37 + 9__ is the same as 95 + 37	45 + 48 is the same as 48 + 4__
1__ + 29 is the same as 29 + 12	68 + 2__ is the same as 22 + 68
59 + 34 is the same as __4 + 59	55 + 45 is the same as 45 + 5__

On the back, explain how you knew how to fill in the missing numbers in each statement.

Flip and Fill

Directions: Players take turns flipping a card and using it to fill a missing number.

9 + 7 + 1 = 9 + ___ + 7	9 + 9 + 1 = 1 + 9 + ___
8 + 5 + 7 = 7 + 5 + ____	3 + 2 + 9 = 9 + ___ + 3
5 + ___ + 5 = 5 + ___ + 9	9 + 6 + 5 = 5 + ___ + 9
4 + 7 + 6 = 4 + ___ + 7	___ + 5 + 8 = 8 + 2 + 5
8 + 9 + ___ = 8 + 2 + 9	3 + ___ + 7 = 7 + 3 + 7
___ + 9 + 5 = 5 + 5 + 9	1 + ___ + 9 = 9 + 1 + 5
3 + 8 + 7 = 7 + ___ + 8	8 + 6 + 4 = 6 + ___ + 8

On the back, explain how you knew how to fill in the missing numbers in each statement.

Flip and Fill

Directions: Players take turns flipping a card and using it to fill a missing number.

16 − ___ = 9 is related to 9 + 7 = 16	___7 − 8 = 9 is related to 9 + 8 = 17
14 − 5 = 9 is related to 9 + ___ = 14	12 − 4 = 8 is related to 8 + 4 = 1___
18 − 9 = ___ is related to 9 + 9 = 18	1___ − 6 = 8 is related to 6 + 8 = 14
12 − 7 = 5 is related to 5 + 7 = 1___	10 − 7 = ___ is related to 3 + 7 = 10
13 − ___ = 5 is related to 5 + 8 = 13	14 − ___ = 6 is related to 6 + 8 = 14
15 − 7 = 8 is related to 8 + ___ = 15	16 − 8 = 8 is related to 8 + ___ = 16
11 − 5 = 6 is related to 6 + ___ = 11	13 − 7 = 6 is related to ___ + 7 = 13

On the back, explain how you knew how to fill in the missing numbers in each statement.

 This resource can be downloaded at **https://qrs.ly/psf6a5o**

ACTIVITY 5.11

Name: Which Switch? (The Center)   **Type:** Center

About the Center: This center is an example of how you can provide follow-up experiences to reinforce ideas taught with other instructional activities. In this activity, students are given a problem and asked to write it two new ways (switching it) using parentheses and/or rearranging the numbers. This center is also a good example of how a practice activity can be a good tool for assessing student understanding as well as their emerging fluency. It's possible that some students will simply place parentheses or move numbers without considering the order in which they want to add them. Conversely, you will find students that group the numbers in the most efficient way as their first choice each time.

Materials: Which Switch? cards, recording sheet

Directions: 1. Students choose a card with three addends from a stack or bag of cards. Note that cards are available to download or you can make your own cards.

2. Students record the problem on the card.

3. Students rewrite the problem two times, each time adapting the original expression by using parentheses to show what they would add first and/or rearranging the numbers to illustrate the order they want to add them in.

4. Students determine which problem is easiest for them to solve and record the sum for that problem.

5. After completing several problems, students choose one of their switches and explain why that switch was easiest for them to solve. (This prompt is included on the recording sheet but not shown in the example below.)

In this example, a student pulled 25 + 7 + 5. They reordered the addends in two different ways. They determined that the last arrangement was easiest for them to find the sum. They shared in their writing (not shown) that they thought this was easiest because they knew 25 and 5 was 30. Then, they simply added 30 and 7.

My Card, Switched, Switched Again

Directions: Choose a card and write down the problem. Then, rewrite the problem in two other ways. Use one of the problems to find the sum most easily.

Problem On My Card	Switched	Switched Again
25 + 7 + 5	7 + 5 + 25	25 + 5 + 7 = 37

RESOURCE(S) FOR THIS ACTIVITY

Which Switch?

Directions: Choose a card and write the problem. Then, rewrite the problem in two other ways using the Associative propperty. Use one of the problems to find the sum most easily.

Problem on My Card	Switched	Switched Again

5 + 7 + 5	25 + 7 + 5	25 + 7 + 25
9 + 8 + 7	39 + 8 + 7	49 + 8 + 17
2 + 9 + 8	2 + 19 + 8	2 + 29 + 28
6 + 8 + 6	36 + 8 + 6	6 + 68 + 46
7 + 5 + 7	7 + 55 + 7	37 + 35 + 7
6 + 9 + 4	6 + 79 + 4	36 + 9 + 14

 Use the Which Switch? cards from Activity 5.3. The Center recording sheet can be downloaded at **https://qrs.ly/psf6a5o**

ACTIVITY 5.12

Name: Triangle Cards   **Type:** Center

About the Center: Triangle cards should be familiar to students. These cards reinforce the inverse relationship between addition and subtraction. In this center, students select a card, examine the numbers, and write the addition and subtraction equations related to the numbers. Templates are available to download. Unlike other cards you may have used, these cards do not circle or signal the sum. The intent is for students to think about how the numbers go together. After students work with cards, ask them to explain how they determined what equations to write. You want them to cite ideas about addition (e.g., "these numbers were combined"), subtraction (e.g., "this number was taken from"), or the relationship between the operations. You don't want notes about the bigger number being the answer (addition) or that you have to subtract from the bigger number (subtraction).

TEACHING TAKEAWAY

Extend games and centers by having students create their own examples. For this center, students can be charged with creating three sets of triangle cards and proving that their cards are accurate.

Materials: Triangle cards, recording sheet (optional)

Directions: 1. Students pull a triangle card.

2. Students record related addition and subtraction equations with the numbers on the card.

3. Optional: Students have to prove that one of their equations is true.

Cards A, B, and C show how you can help students practice the inverse relationship between these operations and see patterns. Note that at first, it makes sense to keep numbers in similar positions (i.e., sum at the top) before moving them all around. After completing the equation sets for these three related cards, you can have students write about how the three cards are related, the patterns they notice in the cards, or create a new card that is related to them (e.g., 40, 34, and 6). Card D wouldn't be used with the other cards. Instead, it is offered as an alternative to three given numbers on a card. With this card, students determine what number the question mark represents and then write the equations associated with those three numbers. For the card shown, a student could determine that the question mark is 6, writing 6 + 30 = 36. But they could also determine it to be 66, writing 30 + 36 = 66.

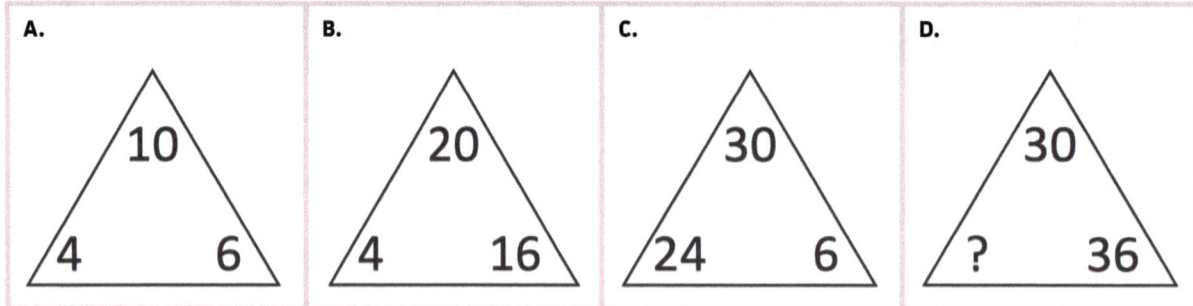

Triangle Cards

Directions: Choose a triangle card. Write the numbers on the card. Use the numbers to write two addition and two subtraction equations.

Card	Addition Equations	Subtraction Equations

On the back: Choose one of your equations and prove that the equation is true with pictures or number lines.

This resource can be downloaded at **https://qrs.ly/psf6a5o**

Properties of Multiplication and the Inverse Relationship With Division

MODULE 6

PROPERTIES OF MULTIPLICATION OVERVIEW

This module is the first of four focused on foundations for multiplication and division. All reasoning strategies are grounded in properties and place value, thus understanding and being able to apply the properties of multiplication are a necessary beginning for students! Properties that are critical to enacting multiplication and division reasoning strategies are listed in the table.

PROPERTY	SYMBOLIC REPRESENTATION	HOW A STUDENT MIGHT EXPLAIN THE PROPERTY
Commutative	$a \times b = b \times a$	"You can multiply numbers in any order and get the same answer."
Associative	$(a \times b) \times c = a \times (b \times c)$	"When you multiply three numbers, you can decide which two to group together to multiply first."
Multiplicative Identity	$a \times 1 = 1 \times a = a$	"When you multiply by one, you get the number you started with."
Inverse Relationship	If $a \times b = c$ then $c \div b = a$ and $c \div a = b$	"In multiplication, the product can be divided by either factor and it will equal the other factor."
Distributive	$a \times (b + c) = a \times b + a \times c$	"You can split one factor into two parts, multiply each part by the other factor, and then add them together."

Source: Adapted from Van de Walle et al., 2023.

Understanding these properties means more than recalling $a \times b = b \times a$ or saying, "The order of factors doesn't matter." Understanding means representing properties and describing why they work. The images here illustrate the commutative and associative properties, respectively.

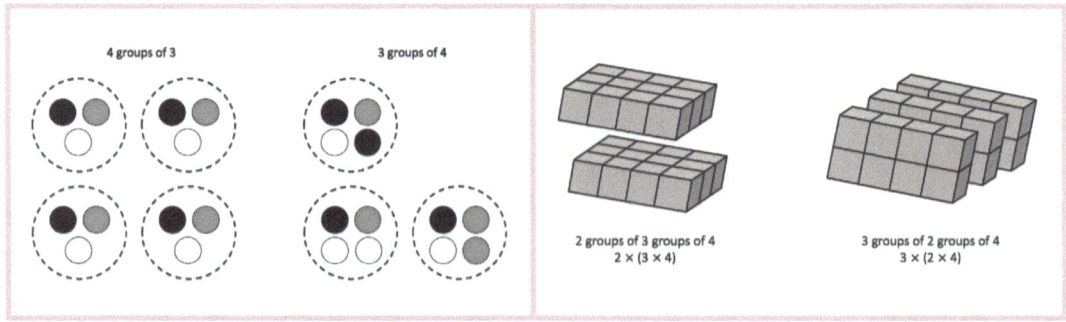

Perhaps the most important property is the distributive property of multiplication over addition. The work on the next page shows the distributive property in action as a student decomposes both factors to multiply on the left and another student decomposes only one of the factors on the right.

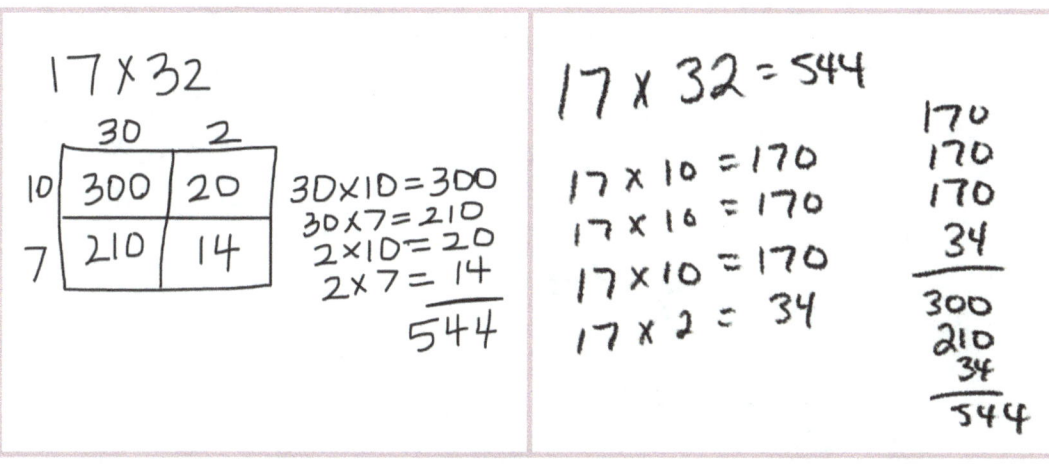

Intuitive understanding of these properties, coupled with a skillful application of them, empowers your students to choose and use efficient strategies as they navigate problems.

And the relationship between multiplication and division is a necessary beginning. The most likely strategy a student will use for division is Think Multiplication, and even the algorithm for division is based on using an inverse relationship; for example, to solve 60 ÷ 3, a student can think, "How many bags of three pencils can I create with the 60 I bought?" (measurement thinking) or they may think, "If I had three bowls for 60 candies, how many could I put in each bowl?" (sharing thinking). In both cases, they can use Think Multiplication to solve (___ × 3 = 60 or 3 × ___ = 60, respectively).

HOW DO THE PROPERTIES OF MULTIPLICATION CONTRIBUTE TO FLUENCY STRATEGIES?

Similar to the properties of addition, the properties of multiplication make reasoning strategies possible. These also allow for other notable efficiencies, such as how we can multiply with multiples of 10 and 100 (Module 7) and reasoning about factors and multiples (Module 8). The use of the distributive property is likely to be the most obvious application to a strategy: when you teach partial products or breaking apart by addends. The distributive property is also used for finding products with compensation. Look at the student's work for 89 × 7. To carry it out, they multiplied 90 × 7 and then took one group of seven away, which symbolically is 7(90 − 1) = 7 × 90 − 7 × 1, which illustrates the use of the distributive property.

$$89 \times 7$$
$$90 \times 7 = 630$$
$$-1 \times 7 = -7$$
$$630 - 7 = 623$$

The associative property is used when students use the Break Apart by Factors strategy (and rearrange the numbers) to solve a problem, such as 25 × 36. The student work belows shows two ways to solve 25 × 36. On the left, the student factors each of the original factors and then re-associates those factors (note that the commutative property is also in action here). On the right, the student re-associates the 4 from 36 to the 25 in order to have the easier expression (100 × 9) to solve.

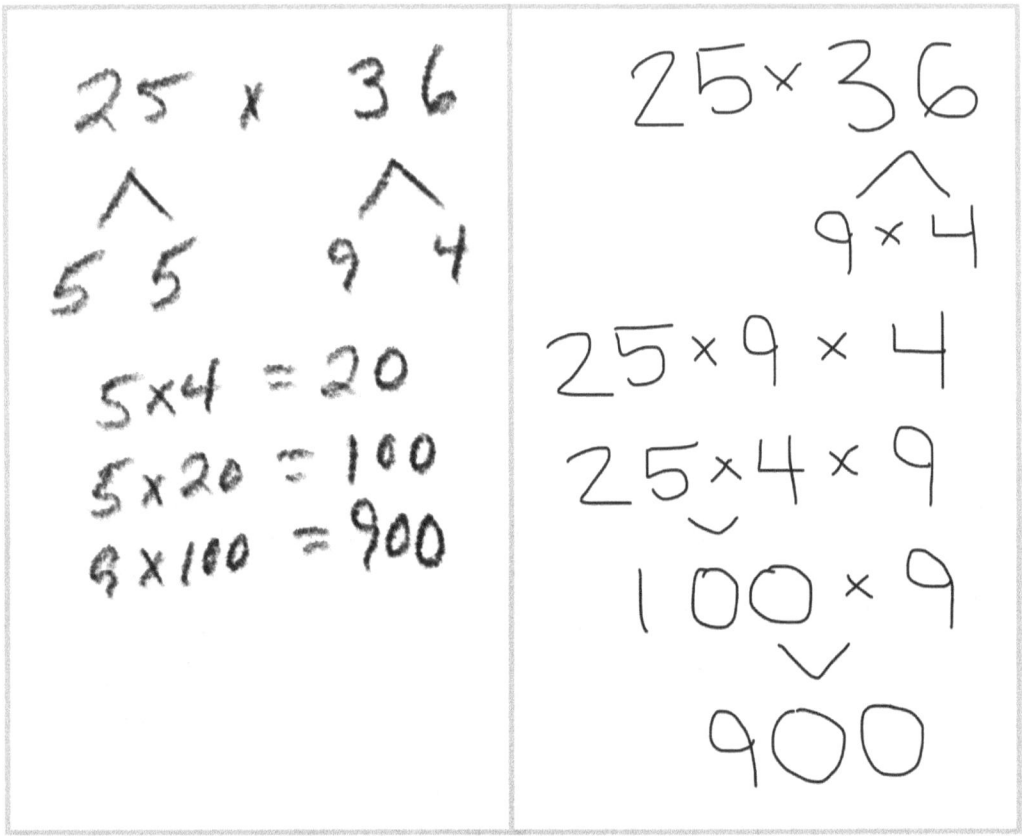

The inverse relationship between multiplication and division is used to find quotients with the Think Multiplication strategy. On the left, the student sees 48 ÷ 6, thinks 6 × 8 = 48, and then translates that to 6 × 80, which we'll examine more of in the next module. On the right, the student sees multiples of five and is able to reason that they can break 535 into 500 and 35 to find a quotient of 107. This will also be unpacked in Module 8.

The properties of multiplication reference page provides an overview, important representations, connections to fluency, and actions to avoid. It can be downloaded and used for reference in planning, teaching, and discussions with colleagues and families.

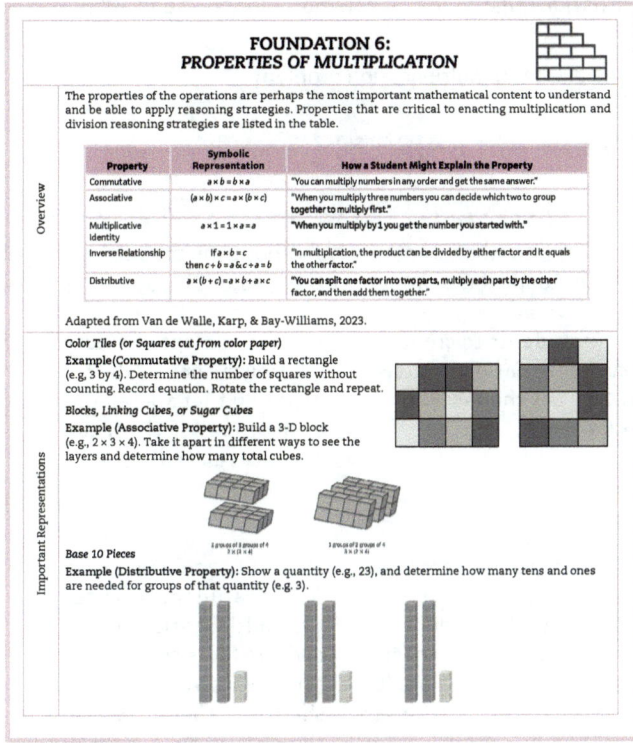

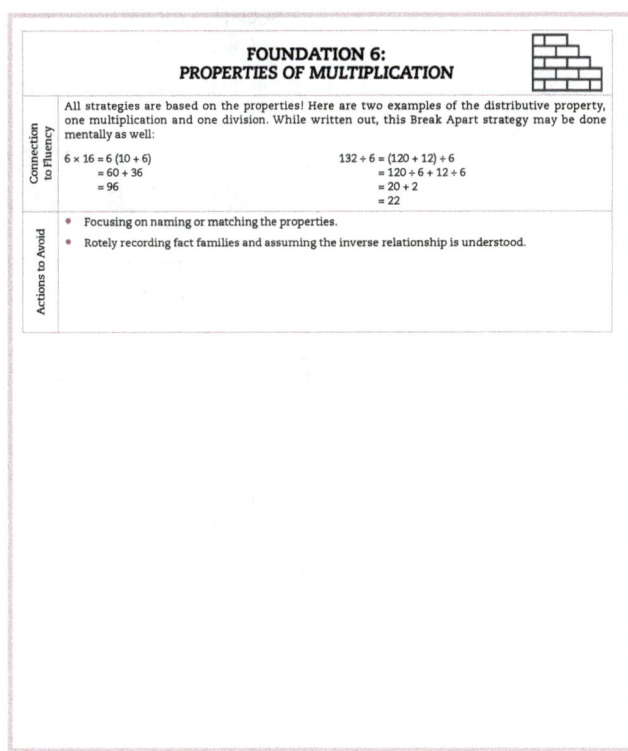

online resources ➤ Available for download at **https://qrs.ly/psf6a5o**

ASSESSING PROPERTIES OF MULTIPLICATION

Assessing properties should not be a simple activity wherein the properties are matched to an example. The key to assessing is to see if students are appropriately *applying* the properties. When you ask students if a multiplication action always works (and ask them to justify their answer), you are likely going to gain strong insights on their understanding of the properties for multiplication.

PROPERTIES OF MULTIPLICATION LOOK-FORS

To begin, attend to students' ability to apply the properties with basic facts. For example, look for students who can recognize commutativity when learning or reviewing their basic facts. For example, to answer 9 × 5, they might think, "It's the same as 5 × 9, which I know equals 45." To answer 49 ÷ 7, a student might say, "I know 7 × 7 is 49, so it's 7." These look-fors with basic facts are an important beginning to look-fors beyond basic facts. This can seem clear with basic fact situations, but as students move to multi-digit problems, they may not recognize opportunities to apply the properties. Look-fors related to the properties and inverse relationships include being able to

- recognize and use commutativity with basic multiplication facts,

- recognize related facts' relationships with basic facts (fact families),

- explain how and why properties work,

- recognize and be able to explain why commutative and associative properties work for multiplication but not for division,
- identify and create examples or visuals of properties,
- rearrange factors to make multiplication more efficient through the commutative and/or associative property (especially in situations that multiply with multiples of 10 and 100, e.g., 4 × 60), and
- solve division problems by recognizing a related multiplication problem.

QUICK ASSESSMENTS FOR PROPERTIES OF MULTIPLICATION

Choose any of these prompts to gain insights into what students know about the properties and to notice the extent that they can apply them. Adapt the prompts to other properties, as appropriate.

Tell the student that 4 × 7 is the same as 7 × 4. Repeat for problem with three factors (e.g., 5 × 7 × 8 and 5 × 8 × 7). Ask them if this is always true (and why). Ask, "How might this idea be useful?"	Ask the students to create a visual or situation to show that 15 × 6 is the same as 10 × 6 + 5 × 6.	Ask the student to create a multiplication and division equation that are related to each other. Ask them how they are related.
Write some equations on index cards (e.g., 12 × 15, 11 × 13, 17 × 7, 26 × 8, 9 × 15, 19 × 3). Ask the student to solve one and ask how they thought about it: For 12 × 15, did they break apart the 12 to solve it (10 fifteens equals 150 and two more equals 180) or did they break apart the 15?	Share that you saw someone solve 8 × 6 by multiplying 8 × 3 × 2. Ask the student if this works (and "Why or why not?"). Note that they don't have to do the calculations to prove that it works. Ask why the student might have used this method.	Share a triangle fact card. Ask the student to write the multiplication and division facts that the card represents. Ask them to share how they know that their equations are correct. Note: Use more than one card and move the location of the product or use something other than a basic fact (e.g., 12, 7, and 84).

EXPLICIT INSTRUCTION FOR PROPERTIES OF MULTIPLICATION AND THE INVERSE RELATIONSHIP WITH DIVISION

Focus your instruction of properties and relationships on the conceptual understanding of these topics. Be sure that students can represent these concepts and their thinking in a variety of ways. During instruction, make sure that they can explain how models and diagrams connect to the properties. As understanding takes hold, shift your focus to generalizing the concepts to larger, more complex numbers. During lessons, be on the lookout for situations where students are challenged to use properties and relationships. A frequent example is when a student has difficulty finding a product or quotient; you intervene by giving a completed example related through the commutative or inverse relationship properties, but they still cannot find the product or quotient.

ACTIVITY 6.1
COMMUTATIVE FRAYER

Developing a deep conceptual understanding comes about through representation (Bruner & Kennedy, 1965). But you have to be careful of singular, preferred representations indicative of representational bias (SanGiovanni et al., 2022). In short, you want your students to represent a concept in a variety of ways and be able to make explicit connections between those representations (Huinker, 2013). A Frayer model is the perfect tool for having students create diverse representations of a concept, such as the commutative property. After creating examples, discuss with students how their representations are similar and different while connecting them explicitly to the expression and the model.

This activity can be done in small groups or in a "four corners" station-rotation model where partners move from station to station completing different prompts together. First, talk with students about the different ways they can represent multiplication. They should come up with equal groups, areas, arrays, number lines, and repeated addition. These become the sections of the Frayer. Feel free to add some possibilities, such as "words that go with it" (lower-right section in the example), "describing it in my own words," or "situation or story." You can give students a choice about which four ideas they use for their Frayer or you can assign topics to them. Identify a multiplication problem for the class to use. Having the same problem will help focus discussions afterward.

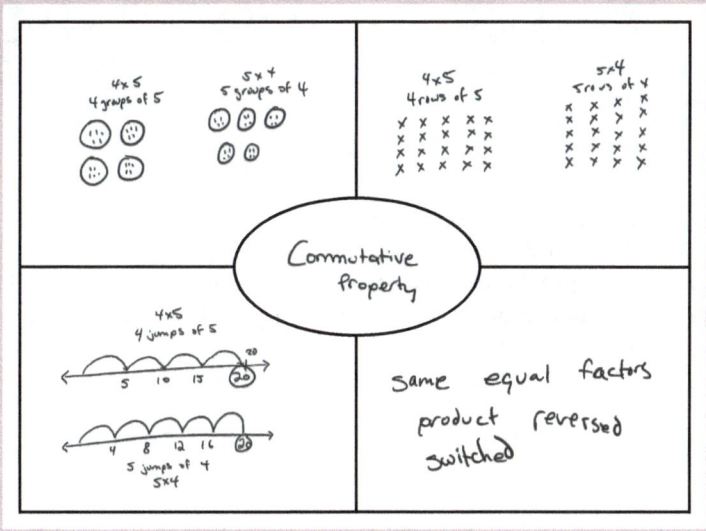

A blank Frayer model is downloadable from **https://qrs.ly/psf6a5o**

This example shows a completed Frayer diagram for 5 × 4. Note that you don't want to use the same factors (e.g., 3 × 3, 6 × 6, etc.) and that you want to keep them somewhat easy to represent. Invite students to compare their representations.

Change the activity to the distributive Frayer! The Frayer graphic works great for a two-digit number times a two-digit number where either one (or both) can be broken apart: For example, take a look at 13 × 21. Rather than show different representations, students illustrate different ways to partition a rectangle to break apart one or both of the factors. Include a section where they add a situation to fit the problem. Students can also complete a Frayer model independently, which is both good practice and a good assessment artifact.

TEACHING TAKEAWAY

Making anchor charts with the class is a powerful way to honor their thinking, help them see themselves in their mathematics class, and position them as doers of mathematics.

ACTIVITY 6.2
THE ASSOCIATIVE PRISM

Rectangular prisms are perfect for exploring the associative property of multiplication. In this instructional activity, students build rectangular prisms and decompose them into smaller prisms to show the associative property. Introduce the exploration with the task on the left below. You want to provide cubes to students to model their work and they should be encouraged to sketch their models, too. The image on the right shows what their models should eventually look like. Essentially, they'll make smaller prisms with the dimensions shown.

With their solutions in place, ask them how the total number of cubes changed (they didn't, there are still 24). Ask them to explain how the prisms are related. Reinforce the expressions of both the original prism and the resulting four prisms. Ask them how they are the same and different.

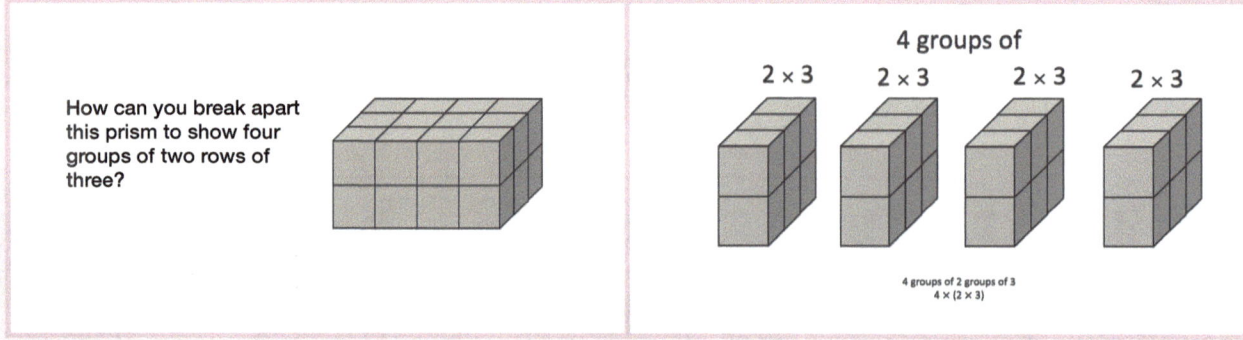

After that first discussion, you can choose to continue the activity in one of two ways (and possibly both). You want them to make an argument about the smaller prisms they created. Simply, you want to ask if the smaller prisms should be described as four groups of two rows of three or two groups of three rows four times (see image). They'll see that the order of the factors change, but the total (product) remains unchanged. Note that this may come up in your debrief of the opening task. Take advantage of that moment if it does.

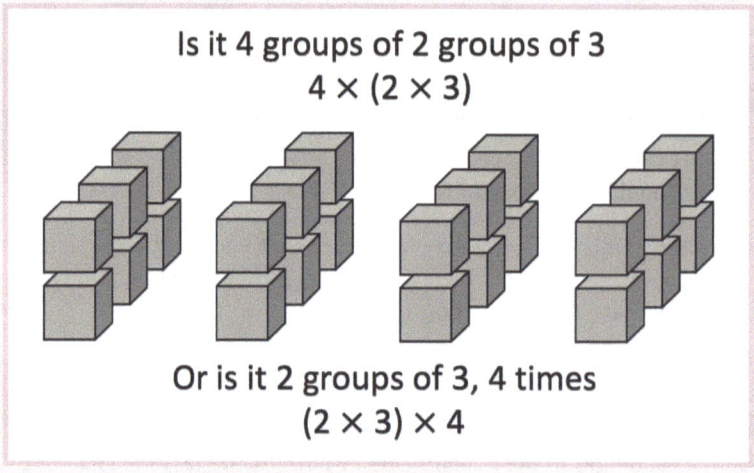

To extend the prompt, have students determine if there are other ways they could decompose the prism of 24 cubes. On the left, you see the extended prompt. It asks about other ways they break apart the prism to show groups of groups. Possibilities are shown on the right. Have them make models and observe how the order of the factors (dimensions) doesn't change the product (volume).

What other ways can you break apart the prism to show groups of groups?

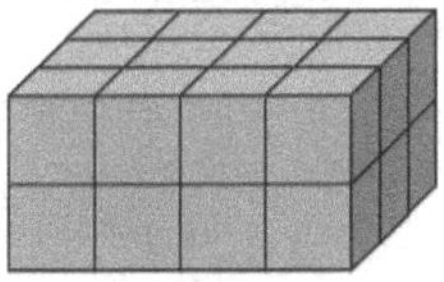

2 groups of 3 groups of 4
2 × (3 × 4)

3 groups of 2 groups of 4
3 × (2 × 4)

After working with a prism of 24 cubic units, you can have students break into smaller groups or rotate through breaking apart other prisms. Good options include prisms of 12, 18, and 30 cubes. Keep in mind that your students do not need to know the formula for volume to work with this task.

ACTIVITY 6.3
YOU HAVE YOUR WORK CUT OUT FOR YOU

This activity focuses on the distributive property using an area model. Students will need 3x 5 index cards, scissors, glue, and a set of problems (these can be from a worksheet, on the board, or you can make cards). To begin, students label their card to represent the problem they are solving. For example, if they are solving 45 × 11, they label the long side of their card with 45 and the short side with 11. Graph paper is a good alternative to index cards because it will show the exact area.

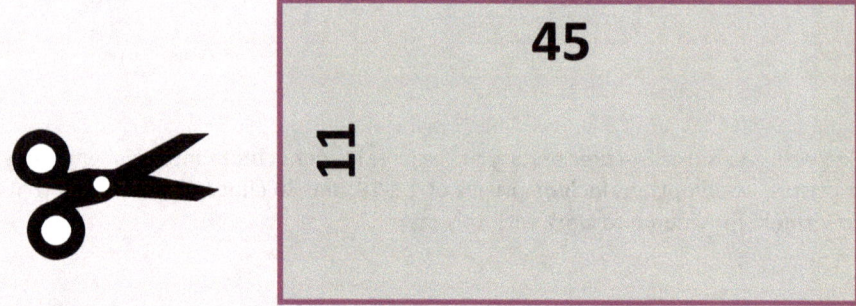

Then, they decide which cut they are going to make. In this problem, they may cut a sliver from the bottom and update the width:

Or they may cut a piece from the end, and update the length:

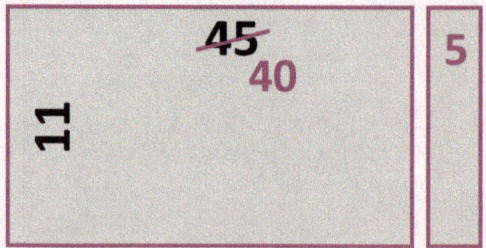

In either case, they glue their partitioned rectangle into their journal or onto a piece of paper and record the original and the new expressions. For example, the first student might record 45 × 10 + 45 × 1 or 45(10+1).

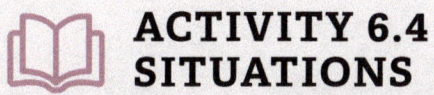 ## ACTIVITY 6.4
SITUATIONS

Contexts for mathematics concepts improve understanding and accessibility of the content. Before this activity, your students should have had ample experience with drawing and representing multiplication and division. In this instructional activity, students will use both to describe given situations. Give a situation card to a group of students. Have them work to represent it with a drawing or diagram and both a multiplication and division equation. After students work on situations, bring the group together to discuss their representations and equations. Be sure that you explicitly connect the written situation to the drawings and equations. You also want to reinforce the relationship between the two operations.

Situation 1	Situation 2
Eighteen pieces of gum are in three packs of six.	Thirty-two pumpkins are planted in four rows of eight pumpkins.
Situation 3	**Situation 4**
Fifty-four people are in six lines of nine.	Les scored 21 points by making seven 3-point baskets.
Situation 5	**Situation 6**
Twenty-eight goldfish are put in seven bowls with four fish each.	Eight rows of five cars make forty cars in the parking lot.
Situation 7	**Situation 8**
A string that is 24 inches is three times longer than a string that is 8 inches long.	Fifty tires are on five dump trucks that have ten tires each.

The student's work shows how the first situation might be represented. Notice that their equal group representations match the problem type. With that in mind, note that Situation 7 is a multiplicative comparison situation that isn't usually introduced until fourth grade.

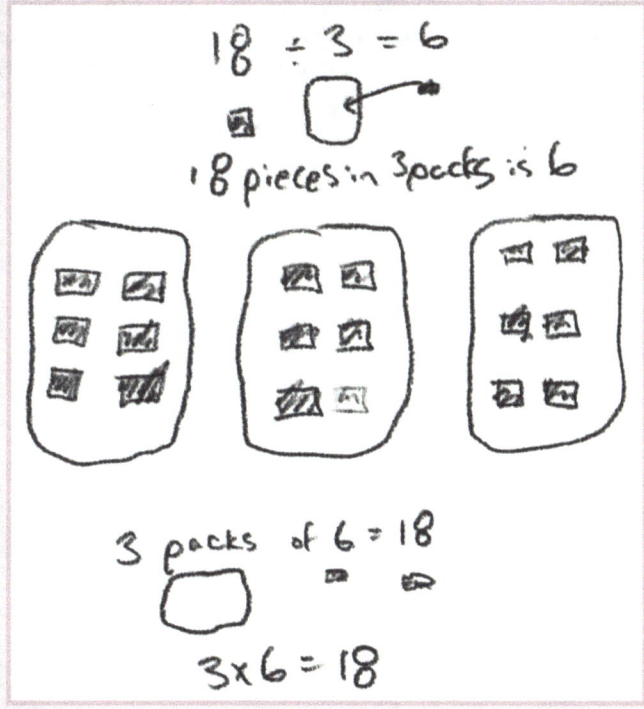

You can use this activity as source material for creating a center. After discussing the different situations, give each student an index card. Have them create a situation (e.g., nine packs of five cards is 45 cards). Collect and review their situations. Label each card with a letter and put them in a shoebox (or something similar). For the center, have students select situation cards, write the letter of the situation, and then show it as both multiplication and division.

ACTIVITY 6.5
WHICH IS IT?

The goal of this activity is for students to see the relationship between multiplication and division. In this experience, group students in pairs or triads. Assign each group two numbers (e.g., 4 and 8). Have them build an array of tiles with the two numbers and describe the array with multiplication, then record the equations. Have them describe the array with division and write those equations. This example shows what students' work might look like for 4 and 8.

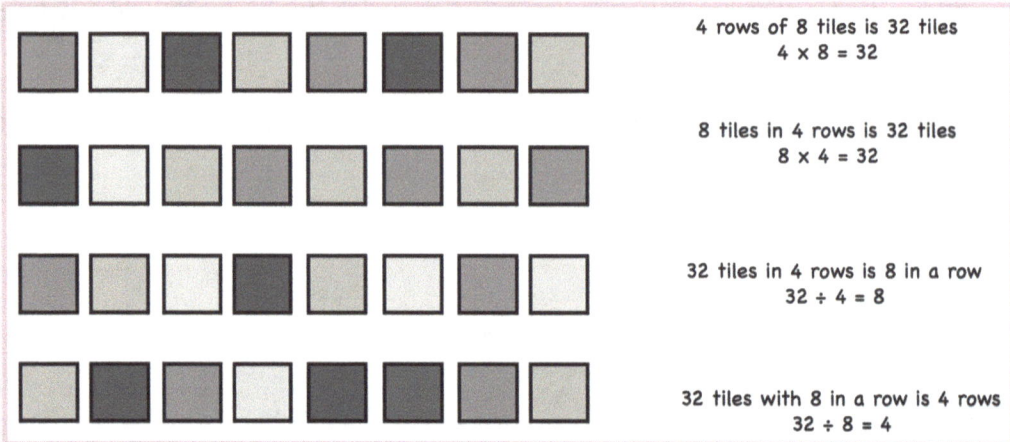

4 rows of 8 tiles is 32 tiles
4 x 8 = 32

8 tiles in 4 rows is 32 tiles
8 x 4 = 32

32 tiles in 4 rows is 8 in a row
32 ÷ 4 = 8

32 tiles with 8 in a row is 4 rows
32 ÷ 8 = 4

This experience shouldn't be your students' first foray into building arrays. You should expect that they are able to build and describe them. Students may decide to build the arrays to find the total number of tiles before pivoting to the division descriptions. You can model an example before students explore a new task with a partner. But if you don't model, you may get a better insight into how well your students really understand arrays and multiplication.

After groups build examples and record their work on chart paper (or something similar), bring the class together to review each other's work. Then, show a new array and ask students if the model shows multiplication or division. Have them talk with partners and hold a final, short discussion about how the model represents both operations.

ACTIVITY 6.6
PROMPTS FOR PROPERTIES
AND INVERSE RELATIONSHIPS

Use the prompts below as opportunities to develop understanding of and reasoning with the strategy. Have students use representations and tools to justify their thinking, including base-ten models, number lines, number charts, and so on. After students work with the prompt(s), bring the class together to exchange ideas. Remember that these could be useful for collecting evidence of student understanding.

- Neil uses the commutative property to rearrange 5 × 6 × 20. How do you think he rearranges the problem? Why do you think he rearranges it? How do you know it will be the same product regardless?

- Use parentheses to show which multiplications you would do first for these problems:

  a. 12 × 5 × 6   b. 6 × 15 × 2 × 50

- Share two ways you could break apart these problems to solve them:

  a. 15 × 9   b. 25 × 14   c. 12 × 104

- Jayden says she can solve division problems by using multiplication. For 72 ÷ 8, she thinks 8 ×? is 72. Why does Jayden's strategy work?

- If you know 6 × 3 = 18, what other related multiplication and division equations do you know?

- Deryn says subtraction goes with addition as division goes with multiplication. What do you think Deryn means by that? Use examples to show your thinking.

- Endy says a rectangle that is 8 inches long and 6 inches wide will have the same area as a rectangle that is 6 inches long and 8 inches wide. Do you agree? Create a model to show your thinking.

- Maisha and Gilly are working on 4 × 2 × 5. Maisha says she thinks of the problem as 8 × 5 = 40 and Gilly says she thinks of it as 4 × 10 = 40. Who do you agree with? Use pictures and drawings to show your thinking.

QUALITY PRACTICE FOR PROPERTIES OF MULTIPLICATION AND THE INVERSE RELATIONSHIP WITH DIVISION

Students need to process what they do, why they do it, why it works, and how it helps. You can help them process in a variety of ways. One way is to add a reflection question at the end of a game or center. Have them complete the question when they're finished. Show that you value that reflection just as much as how well they stay on task, how well they play with a partner, or how correct their work is. Alternatively, you can hold brief whole-group discussions about the games and centers they've been working with over the course of a few days.

ACTIVITY 6.7

Name: "The Truth"   **Type:** Routine

About the Routine: Students may seem to understand properties of multiplication and its inverse relationship to division on some practice sheets. At other times, when needing to call on or apply that understanding to a new situation, it seems as though they've never heard of it before. They may not fully understand the relationship or how to use it because they have simply moved numbers around during practice with little to no thought. This routine aims to develop a deeper, accurate understanding causing them to pause, consider, and explain.

Materials: This routine does not require any materials.

Directions: 1. Pose three pairs of equations. Give students individual think time to decide which are true and which are false.

2. Have students discuss if each pair of equations are related through the inverse relationship (top example) or a property of multiplication (bottom examples).

3. After students discuss with partners, bring the group together to share ideas. When students identify equations that aren't related, have them determine a new equation that would be a true relationship to either of the equations.

These are different examples of the routine. You can provide equations with all numbers provided or with an unknown. It is wise to use numbers/facts that may be less familiar to students, causing them to truly grapple with the prompts. You can also rely on problems that aren't basic facts (e.g., $12 \times 3 = 36$ or $25 \times 4 = 100$). The commentary to the right is not part of the prompt to students; it is provided as an example of the reasoning students might use and, in turn, will help you think about the examples you might create.

Inverse Relationship

WHICH IS THE TRUTH?	EXAMPLE OF STUDENT REASONING
A. $8 = 7 \times 56$ is the inverse of $56 \div 7 = 8$	This is false, but students might think it is true because they can muddy ideas about the order of numbers being reversed and the inverse relationship of these operations.
B. $6 \times 8 = 48$ is the inverse of $6 = 8 \div 48$	This is false. The order of the symbols has been reversed with the new operation.
C. $63 \div ? = 9$ is the inverse of $9 \times ? = 63$	This is the true statement. However, the factors appear in a different order, which may trip up some students.

Commutative Property

WHICH IS THE TRUTH?	EXAMPLE OF STUDENT REASONING
A. $7 \times 5 = 35$ and $35 \times 5 = 7$ show the commutative property	This is false; it reverses the order of the numbers without regard to the operation.
B. $5 \times 9 = 45$ and $45 = 9 \times 5$ show the commutative property	This truth could be tricky for some because the product is reversed. It's also a true example of the property, as the order of the factors is reversed.
C. $54 = 6 \times 9$ and $54 = 9 \times 6$ shows the commutative property	This is a true statement with the product on the left side of the equal sign. (It is included to show that you can have any number of truths or lies in this routine.)

Associative Property

WHICH IS THE TRUTH?	EXAMPLE OF STUDENT REASONING
A. (6 × 7) × 8 and (8 × 7) × 6 show the associative property	This is a true example of the associative property. It reverses the order of the factors, and students might think of this as the commutative property.
B. (12 × 3) × 9 and (3 × 21) × 9 show the associative property	This is false; it changes the order of some factors (3 and 9) while changing the order of digits in the third factor (12 to 21).
C. (2 × 3) × (5 × 2) and (3 × 5) × (2 × 2) show the associative property	This truth has four factors. Students may assume that the associative property is about the order of three factors.

Distributive Property

WHICH IS THE TRUTH?	EXAMPLE OF STUDENT REASONING
A. 18 × 9 = 9 × 2 × 9 shows the distributive property	This is false. The equation is correct, but it is not distributive property because the 18 wasn't broken into an addition or subtraction expression.
B. 125 × 6 = 100 × 6 + 20 × 6 + 5 × 6 shows the distributive property	This is true because 125 was broken apart into 100 + 20 + 5 and each part was multiplied by 6.
C. 27 × 9 = 27 × 10 − 27 shows the distributive property	This is true because 9 was changed to 10 − 1 and both were multiplied by 27.

ACTIVITY 6.8

Name: Multiplication Match

Type: Game

About the Game: Students need to be able to apply the commutative and associative properties when multiplying three or more numbers. When a student is faced with 5 × 3 × 6, it most likely will be easier for them to rearrange the order of the factors to 5 × 6 × 3 and multiply. Finding three groups of 30 is easier for most students than fifteen groups of 6. This game will help students recognize problems that contain the same numbers; however, the numbers are listed in a different order. Students will play a game similar to Memory, where they find pairs of cards that match. Matching pairs will have expressions that contain the same numbers but the order of these numbers is switched as a result of applying the commutative and associative properties of multiplication.

Materials: *Multiplication Match* cards

Directions:
1. Two students lay a set of 24 cards face down in a 6 × 4 array.

2. Player 1 turns over two cards. If the cards do not have expressions that match, the student turns the cards back over to their original spot in the array and the player loses their turn. If the cards do have matching expressions, the player says, "It's a Multiplication Match!" Once a player finds a match, they select one of the two options to solve. If their answer is correct, they keep the cards and get another turn. If a player says an incorrect answer, the cards are returned to the array and play goes to the next player.

3. Players take turns until all cards are removed from the array. The player with the most cards at the end of the game is the "Multiplication Match" winner.

4. Cards can be combined, shuffled, and put back in a new 6 × 4 array to play again.

For example, a player selects cards 3 × 5 and 7 × 3; the cards are turned face down again and it is the next player's turn. If a 2 × 4 × 6 and 6 × 4 × 2 is revealed, the player gets to keep the cards if they determine the product to be 48. Then that player turns over two new cards. The images show the different sets of cards that can be downloaded. The set on the left includes multiplication expressions where only the commutative property could be applied, while the right contains expressions where both the commutative and associative properties could be used.

RESOURCE(S) FOR THIS ACTIVITY

Multiplication Match (Set 1)		
5 × 4	4 × 5	3 × 6
6 × 3	7 × 2	2 × 7
8 × 2	2 × 8	7 × 3
3 × 7	6 × 4	4 × 6

Multiplication Match (Set 3)		
7 × 3 × 5	7 × 5 × 3	5 × 6 × 4
6 × 5 × 4	5 × 3 × 4	3 × 5 × 4
2 × 2 × 7	2 × 7 × 2	3 × 2 × 9
3 × 9 × 2	5 × 7 × 6	5 × 6 × 7

online resources ⤷ This resource can be downloaded at **https://qrs.ly/psf6a5o**

ACTIVITY 6.9

Name: Triangle Card Cover   **Type:** Game

About the Game: *Triangle Card Cover* comes from triangle fact cards that model the relationships between multiplication and division. In this game, players generate a number to complete a triangle with the goal of being the first player to complete all of the triangles on their board. As students play, have them voice two of the equations their triangle represents. For example, a player who rolls a 3 to complete a triangle of 3, 9, and 27 would say, "3 times 9 equals 27 and 27 divided by 3 equals 9."

Materials: Playing cards (queens = 0, aces = 1; remove tens, kings, and jacks) or a 10-sided die, *Triangle Card Cover* game boards, chips or counters for game pieces

Directions: 1. Players take turns generating a digit with dice or cards.

2. If the digit completes a triangle card, the player covers it with a chip.

3. If the digit can't be used, the player loses their turn.

4. The first player to cover all of their triangles wins.

Two game boards are available to download. You can use that file to make additional game boards by duplicating a game board and changing some of the numbers. This works well for holding a game with three or four players instead of only two.

> **TEACHING TAKEAWAY**
>
> Have students voice their thinking to help them process it and have others give feedback. This extra step slows processing, reinforces precision, and provides an element of accountability.

RESOURCE(S) FOR THIS ACTIVITY

Triangle Card Cover

Directions: Take turns rolling numbers. Use the number your roll to complete a triangle card. The first player to complete all of the triangles wins.

| 8 | 27 | | 9 |
| 32 | 3 | 4 — 16 | 45 |

| | 72 | 6 | |
| 3 — 18 | 8 | 42 | 21 — 7 |

| 30 | 7 | 9 | 7 |
| 6 | 63 | 18 | 56 |

Triangle Card Cover

Directions: Take turns rolling numbers. Use the number your roll to complete a triangle card. The first player to complete all of the triangles wins.

| 8 | 18 | | 9 |
| 48 | 3 | 4 — 20 | 54 |

| | 72 | 7 | |
| 3 — 21 | 9 | 42 | 28 — 4 |

| 36 | 7 | 8 | 8 |
| 6 | 42 | 16 | 64 |

online resources — This resource can be downloaded at **https://qrs.ly/psf6a5o**

ACTIVITY 6.10

Name: Property Math Libs   **Type:** Center

About the Center: This center is an easy way for students to generate numbers and complete statements about the associative or distributive property. Students are challenged to complete all the lines in the fewest number of cards possible. Afterward, students reflect on which example is the best use of the associative property. You can swap out the numbers to create different experiences.

Materials: Playing cards (queens = 0, aces = 1; remove tens, kings, and jacks), digit cards, or 10-sided dice; recording sheet

Directions:

1. Students flip a card and use the digit to fill in one of the blanks on their recording sheet.

2. Once both blanks are filled for a line item, the student records the product.

3. The goal of the center is to complete all blanks in as few cards as possible.

4. After completing the game, the student determines which example was a good use of the property and why.

RESOURCE(S) FOR THIS ACTIVITY

Associative Property Math Libs

Directions: Flip a card. Use the digit to complete the statement. When you complete a statement, write the product. Try to complete all of the blanks in the fewest cards.

(___ × 7) × 10   is the same as   (10 × 3) × ___   = _____

(5 × ___) × 2   is the same as   (5 × ___) × 9   = _____

(___ × 3) × 4   is the same as   (___ × 4) × 6   = _____

(4 × ___) × 5   is the same as   (5 × 5) × ___   = _____

(___ × 7) × 5   is the same as   (2 × 5) × ___   = _____

(3 × ___) × 8   is the same as   (___ × 5) × 3   = _____

(7 × ___) × 5   is the same as   (6 × 5) × ___   = _____

Which use of the associative property made it easier for you to multiply? Why?

Distributive Property Math Libs

Directions: Flip a card. Use the digit to complete the statement. When you complete a statement, write the product. Try to complete all of the blanks in the fewest cards.

2___ × 7   is the same as   20 × ___ + 8 × 7   = _____

45 × 8   is the same as   40 × 8 + 5 × ___   = _____

1___ × 3   is the same as   10 × 3 + 6 × 3   = _____

43 × 5   is the same as   40 × 5 + ___ × 5   = _____

6___ × 4   is the same as   60 × 4 + 7 × 4   = _____

8___ × 5   is the same as   ___0 × 5 + 3 × 5   = _____

47 × 9   is the same as   40 × ___ + 7 × ___   = _____

Which use of the distributive property made it easier for you to multiply? Why?

online resources → This resource can be downloaded at **https://qrs.ly/psf6a5o**

Multiplying by Tens and Hundreds

MULTIPLYING BY TENS AND HUNDREDS OVERVIEW

Multiplying by tens and hundreds (e.g., 3 × 10 and 3 × 100), including multiples of tens (3 × 60) and hundreds (3 × 600), blends conceptual understanding, basic fact automaticity, and the use of properties of multiplication. It appears as a curriculum standard most often as a singular topic in Grade 3. Successful teaching and learning of multiplying by tens and hundreds requires a conceptual focus, helping students understand why zeros present in the factors become zeros in the answers. This is a necessary foundation for multiplying and dividing multi-digit numbers in later grades, simplifying fractions, working with ratios, and many other things.

Multiplying by 100 does *not* mean "Add two zeros to your answer." That is mathematically wrong and therefore unprofessional to use as an explanation. If we simply apply the foundational properties from the previous chapter, we can readily make sense of this equation:

$$20 \times 40 = 2 \times 10 \times 4 \times 10$$

Here, we have simply used our place-value understanding or our multiples of ten knowledge. Applying the distributive property, the problem can be reorganized to $2 \times 4 \times 10 \times 10$, which equals 8×100 or 800. Using this same idea in division, fraction notation is helpful in showing this relationship:

$4800 \div 80$ is the same as $\dfrac{4,800}{80}$

$$\begin{aligned}\frac{4,800}{80} &= \frac{48 \times 100}{8 \times 10} \\ &= \frac{48}{8} \times \frac{100}{10} \\ &= 6 \times 10 \\ &= 60\end{aligned}$$

While this looks like a lot of steps, the idea is that the multiples of ten are factored out, divided, and then multiplied back in once the basic fact has been computed. This module helps students understand why multiplication looks like zeros are getting moved to an answer and why division looks like zeros are being canceled.

HOW DOES MULTIPLYING BY TENS AND HUNDREDS CONTRIBUTE TO FLUENCY STRATEGIES?

Try to create a multi-digit multiplication problem that doesn't call for multiplying with tens or hundreds. You can't! It's impossible because this is inseparable from multi-digit whole-number multiplication and division. Finding these products is a component of every strategy, algorithm, and approach to determining reasonableness, as illustrated by these three great examples. On the left, the student uses the Break Apart by Addends (partial products) strategy to find 38 × 7. It begins with finding 30 × 7. In the center example, the student solves 28 × 27 with a partial products algorithm and the standard algorithm. Both require products of ten. On the right, another student finds the product of 42 × 19. But first, they estimate the product to be about 800, thinking 40 × 20 (not shown in the work). From there, you see their partials play out, again finding products of ten.

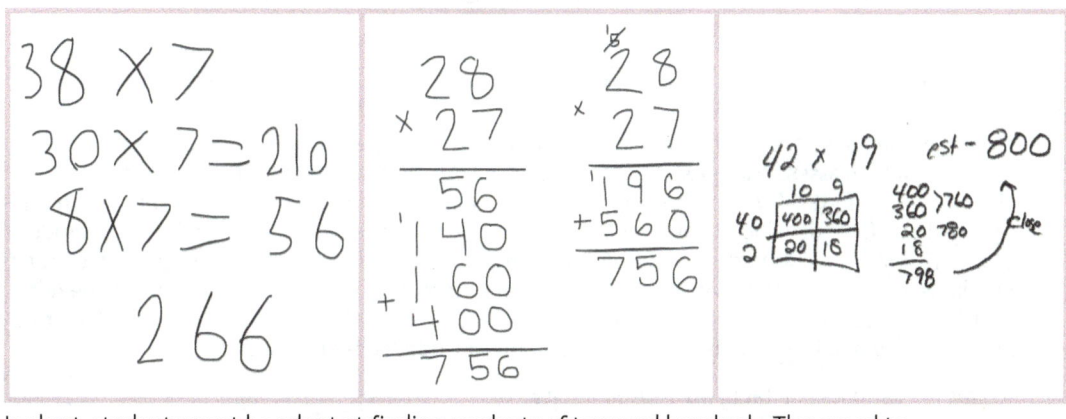

In short, students must be adept at finding products of tens and hundreds. They need to understand the concept and make connections to basic facts.

The multiplying by tens and hundreds reference page provides an overview, important representations, connections to fluency, and actions to avoid. It can be downloaded and used for reference in planning, teaching, and discussions with colleagues and families.

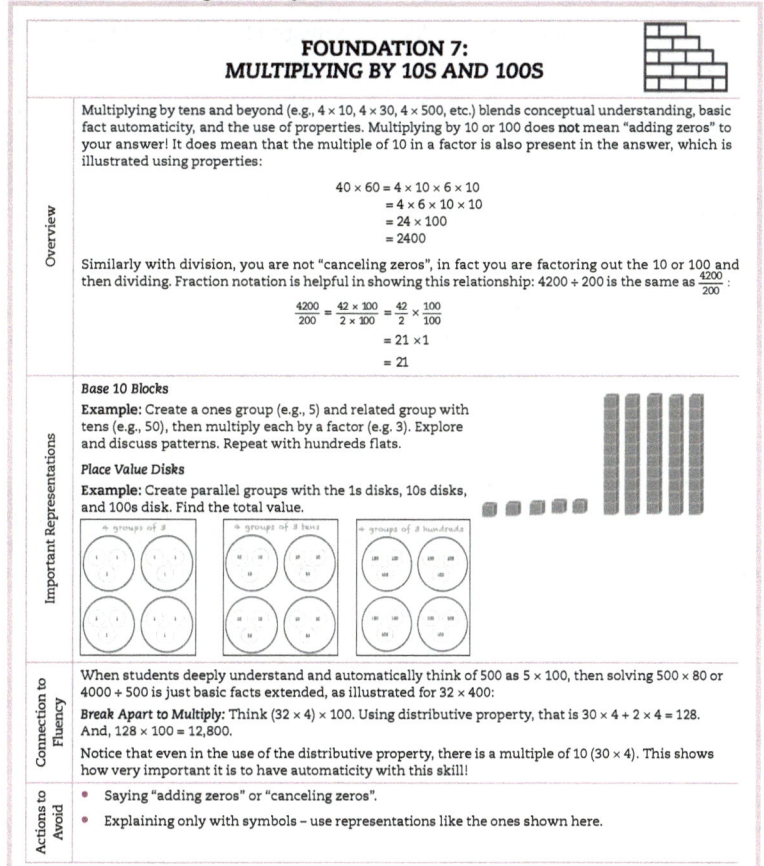

<table>
<tr><td colspan="2">FOUNDATION 7:
MULTIPLYING BY 10S AND 100S</td></tr>
<tr><td>Overview</td><td>Multiplying by tens and beyond (e.g., 4 × 10, 4 × 30, 4 × 500, etc.) blends conceptual understanding, basic fact automaticity, and the use of properties. Multiplying by 10 or 100 does not mean "adding zeros" to your answer! It does mean that the multiple of 10 in a factor is also present in the answer, which is illustrated using properties:

$40 \times 60 = 4 \times 10 \times 6 \times 10$
$= 4 \times 6 \times 10 \times 10$
$= 24 \times 100$
$= 2400$

Similarly with division, you are not "canceling zeros", in fact you are factoring out the 10 or 100 and then dividing. Fraction notation is helpful in showing this relationship: $4200 \div 200$ is the same as $\frac{4200}{200}$:

$\frac{4200}{200} = \frac{42 \times 100}{2 \times 100} = \frac{42}{2} \times \frac{100}{100}$
$= 21 \times 1$
$= 21$</td></tr>
<tr><td>Important Representations</td><td>*Base 10 Blocks*
Example: Create a ones group (e.g., 5) and related group with tens (e.g., 50), then multiply each by a factor (e.g. 3). Explore and discuss patterns. Repeat with hundreds flats.

Place Value Disks
Example: Create parallel groups with the 1s disks, 10s disks, and 100s disk. Find the total value.</td></tr>
<tr><td>Connection to Fluency</td><td>When students deeply understand and automatically think of 500 as 5 × 100, then solving 500 × 80 or 4000 ÷ 500 is just basic facts extended, as illustrated for 32 × 400:

Break Apart to Multiply: Think (32 × 4) × 100. Using distributive property, that is 30 × 4 + 2 × 4 = 128. And, 128 × 100 = 12,800.

Notice that even in the use of the distributive property, there is a multiple of 10 (30 × 4). This shows how very important it is to have automaticity with this skill!</td></tr>
<tr><td>Actions to Avoid</td><td>• Saying "adding zeros" or "canceling zeros".
• Explaining only with symbols – use representations like the ones shown here.</td></tr>
</table>

online resources — Available for download at **https://online qrs.ly/psf6a5o**

ASSESSING MULTIPLYING BY TENS AND HUNDREDS

Assessing this foundation must attend to whether students understand the impact of multiplying by tens. It is not enough (and incorrect) to say, "They are adding a zero to the number." Attending to precise language is important as you look for and listen to students' understanding of this foundation.

MULTIPLYING BY TENS AND HUNDREDS LOOK-FORS

Finding products of tens and hundreds is linked to automaticity with multiplication facts. So, it's reasonable to assess students' progress toward automaticity with multiplication facts. Understanding and skill with multiplying by tens and hundreds should grow as students' automaticity with basic facts develops. Look to see if a student extends their knowledge of 7×3 to a product of tens (7×30) and hundreds (7×300). In time, you should see them blend multiplying by tens and hundreds with ideas of the inverse relationship of multiplication and division, recognizing the relationship between $45 \div 5$ and related expressions such as $450 \div 5$, $4500 \div 5$, and $450 \div 50$. Keep this in mind as you monitor students' progress. Look-fors include students being able to

- recall multiplication facts with automaticity,
- recall division facts with automaticity,
- find a quotient by recognizing the related multiplication fact,
- show and explain that $10 \times 10 = 100$,
- factor a multiple of 10 (e.g., 80) into the factors (e.g., 8×10),
- explain how multiplying by single-digit factors is related to similar problems with factors that are multiples of 10 and 100, and
- use basic fact knowledge to readily solve problems that include a basic fact (e.g., for 20×6, a student will think $2 \times 6 = 12$ and then 10 times that is 120).

QUICK ASSESSMENTS FOR MULTIPLYING BY TENS AND HUNDREDS

The prompts below can be used to determine the extent that students understand and are able to multiply by multiples of tens and hundreds. Select one or more, and pose them in a one-on-one interview, or ask students to prepare a written explanation. Additional prompts can be found in Activity 7.5.

Write $7 \times 8 = 56$ and show it to the student. Ask them to find 7×80 and 7×800. Have them explain how they found the answers.	Write $6 \times 4 = 24$ and $3 \times 8 = 24$. Show these equations to the student. Be clear that the product of both is 24. Ask if they can use either equation to determine 80×3. Have them explain their thinking. Then have them create an equation that they could use 6×4 to help them solve.	Show $3,500 \div 7$. Ask the student what basic fact or facts could help them figure out the quotient. Have them show and explain their thinking.
Ask how multiplying by tens and hundreds is related to basic facts. Ask the student to use examples to explain their thinking.	Share that someone thinks 3×5, 3×50, and 3×500 are all the same thing. Ask the student to explain why they agree or disagree with the statement.	Share that $4 \times 8 = 32$ can help you find the answer to 40×8 and a few other problems. Show the student $40 \times 8 = 320$, then ask what some other problems this might help with. Have them explain their thinking.

EXPLICIT INSTRUCTION FOR MULTIPLYING BY MULTIPLES OF TENS AND HUNDREDS

As noted throughout this book, explicit instruction begins with concrete visuals and stories. For multiplying by tens and hundreds, a good place to start is with base-ten blocks, a proportional model that shows tens and ones. Cuisenaire Rods also show tens and ones proportionally. Eventually, place-value disks can also illuminate the relationships between expressions such as 8×2, 8×20, and 8×200.

TEACHING TAKEAWAY

While some instructional activities are described with a specific representation, you can swap out one representation for another to help students better understand mathematics. What's most important is to explicitly connect the factors and products in the equations with what is represented by the models.

NOTES

_____

_____

_____

_____

_____

_____

_____

_____

_____

_____

_____

_____

ACTIVITY 7.1
MULTIPLES OF TENS AND HUNDREDS

In this activity, students use their knowledge of basic facts and the associative property to multiply by multiples of 10 (and later 100). Using base-ten blocks is one way to help students see the multiples of 10. First, pose a basic fact expression, such as 3 × 4. Briefly discuss the product before posing 3 × 40. Provide base-ten blocks so that students can create representations of both.

3 × 4			
Thousands	Hundreds	Tens	Ones
			▯▯▯ ▯▯▯ ▯▯▯ ▯▯▯

3 × 40			
Thousands	Hundreds	Tens	Ones
		▮▮▮▮ ▮▮▮▮ ▮▮▮	

After they make models of these problems, record the expressions for each problem, making the connections explicit. For example, 3 × 40 can be thought of as 3 × 4 tens; 40 can be thought of as 4 × 10, and so 3 × 40 is the same as 3 × 4 × 10 or 12 × 10 (which is also shown with the base-ten blocks).

$$3 \times 40 \qquad 3 \times 40 \qquad 3 \times 4 \times 10$$
$$4 \times 10 \qquad 3 \times 4 \times 10 \qquad 12 \times 10$$
$$120$$

As the numbers get larger, base-ten blocks are not efficient. For example, it would be difficult to use base-ten blocks to solve 30 × 40. This is why the connections to expressions (as shown) and the deep understanding of break-apart factors are so important. To solve 30 × 40, the factors can be decomposed into 3 × 10 × 4 × 10 and then rearranged to become 3 × 4 × 10 × 10 and eventually 12 × 100. This work is the underpinning as to why it looks as though we are adding zeros when multiplying whole numbers; for instance, 3 × 4 is 12, which is multiplied by 10 and by 10 again.

There are patterns when multiplying and dividing by multiples of 10. Those have to do with the number of zeros in the product or quotient. Students need to be aware of the patterns and why they exist; it is not appropriate to say, "Multiply and add zeros." This shortcut not only denies students an opportunity to understand the impact of multiplying by a power of 10, but it also will not work with decimal values.

ACTIVITY 7.2
GROUPS OF DISKS

Students' first experiences with multiplication are centered on multiplication as "groups of." This instructional activity aims to help students revisit these experiences while establishing a connection between single-digit multiplication and related problems that multiply by multiples of tens and hundreds. This activity uses place-value disks and the Groups of Disks recording sheet (optional).

The task is designed for stations where students rotate through each station to create representations of different multiplication problems (five to eight stations works well). Each station needs place-value disks and a set of related expressions. At each station, students make a physical model with place-value disks before drawing a diagram and writing the equation. The image shows a set of expressions (upper-left) and the corresponding student diagrams for a station with 4 × 3, 4 × 30, and 4 × 300.

TEACHING TAKEAWAY

If you don't have place-value disks in your school, you can create them by writing on counters or poker chips with Sharpie markers or by printing the paper examples from the online resources for this book.

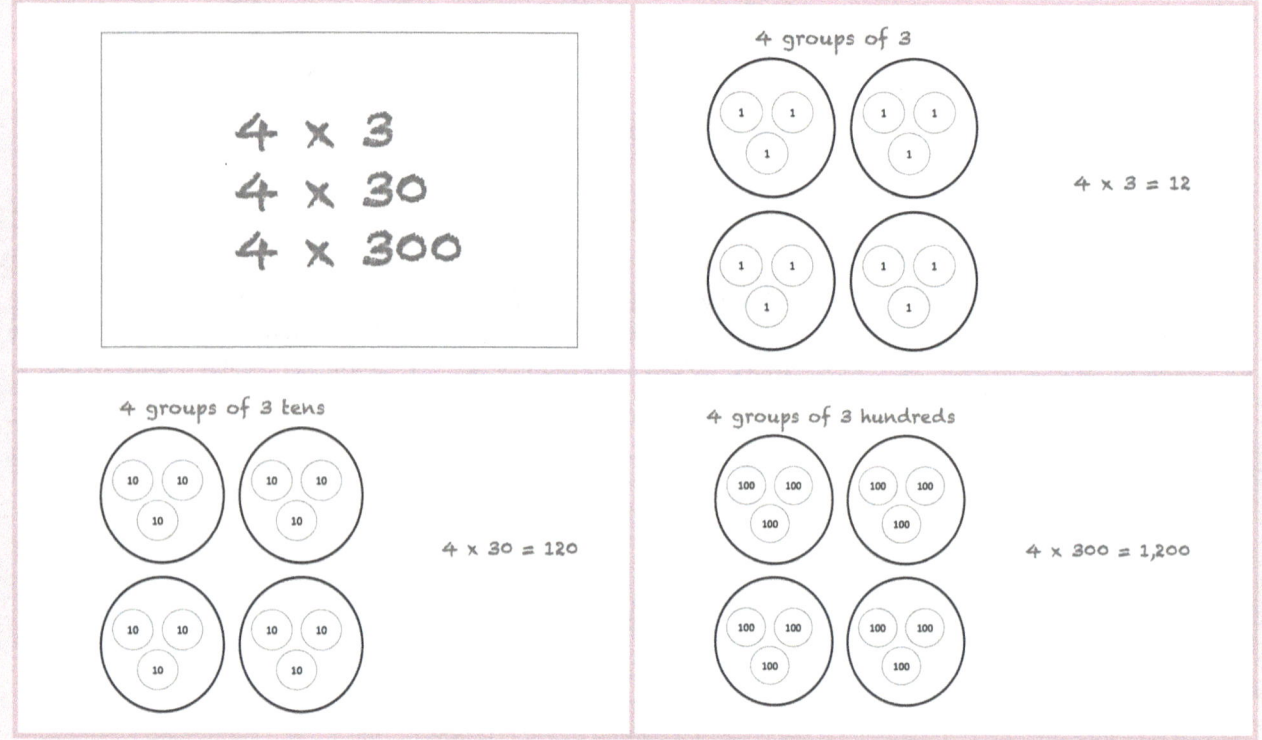

The images show what one of the stations would look like. Each station has a different set of related expressions (or you can have eight stations with four distinct tasks, rotating students through four stations). After students circulate through the stations, bring them together for discussion. First, focus on one set of expressions, such as 4 × 3, 4 × 30, and 4 × 300. Have students compare their representations. Ask them how the problems are similar and how they are different. Ask them what they notice about the factors and products in each equation.

Next, record the other equations from the different stations. Arrange them on the board so that a set of related expressions are aligned vertically and that factors with ones, tens, and hundreds are lined up horizontally, as shown below.

4 × 3 = 12	5 × 6 = 30	2 × 7 = 14	8 × 3 = 24	9 × 3 = 27
4 × 30 = 120	5 × 60 = 300	2 × 70 = 140	8 × 30 = 240	9 × 30 = 270
4 × 300 = 1,200	5 × 600 = 3,000	2 × 700 = 1,400	8 × 300 = 2,400	9 × 300 = 2,700

This arrangement should help students see patterns. Ask them what conclusions they think they can make based on the equations and why they are true. After students share, ask them to create new problems that demonstrate similar patterns.

You can extend the activity by asking students to make models of a problem such as 2 × 4, 20 × 4, and 200 × 4. However, you want to be sure there aren't enough disks to make 20 groups of four or 200 groups of four. When students say they can't make those models because there aren't enough disks, ask them how they could think differently about the problems. They could, for example, think of twenty groups of four as four groups of 20 or four groups of two tens. This extension helps by reinforcing the utility of the commutative property. Additionally, this activity can be adapted to division; for example, students can explore 8 ÷ 2, 80 ÷ 2, 800 ÷ 2, 80 ÷ 20, and 800 ÷ 20.

RESOURCE(S) FOR THIS ACTIVITY

This resource can be downloaded at **https://qrs.ly/psf6a5o**

ACTIVITY 7.3
MULTIPLYING TENS WITH CUISENAIRE RODS

Cuisenaire rods are perfect for all sorts of mathematical explorations and experiences, including finding products of expressions with multiples of ten. That's because there are ten rods in a set. Each rod is 1 centimeter or 10 millimeters longer than the next. Students can line up a number of rods of the same color and measure the whole length to create a physical model of a multiplication problem. In the example, four blue rods were lined up. Each rod is 9 centimeters long, so the length is 36 centimeters. An equation to describe the model is written below. The student measures with a meter stick or ruler to confirm their accuracy. They then find the length of the rods in millimeters. As you know, the length (or product) will be ten times longer because there are 10 millimeters in a centimeter.

TEACHING TAKEAWAY

Find things that are a meter or longer, such as a cafeteria table or bookcase, to have students explore multiplying by hundreds.

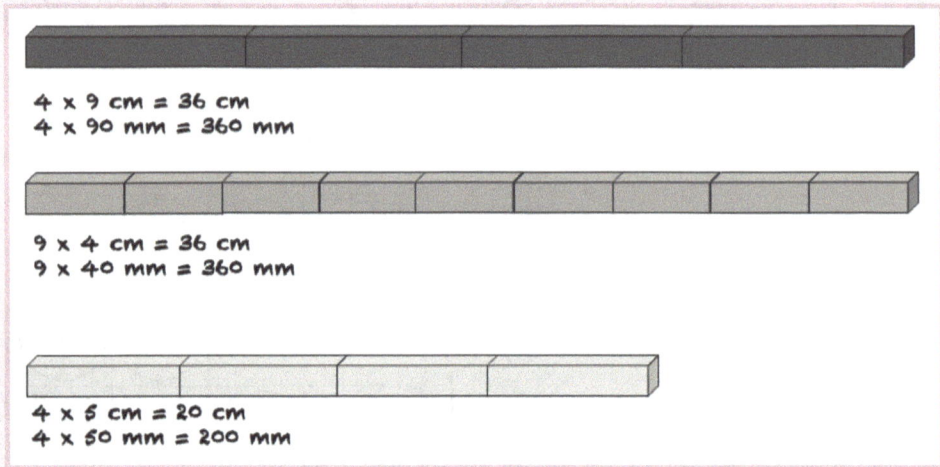

4 x 9 cm = 36 cm
4 x 90 mm = 360 mm

9 x 4 cm = 36 cm
9 x 40 mm = 360 mm

4 x 5 cm = 20 cm
4 x 50 mm = 200 mm

With this experience, you can have students examine different patterns when multiplying by ones and tens. The first (and most obvious) pattern is that of multiplying with tens. For example, 4 × 5 = 20 and 4 × 50 (5 tens) = 200 (20 tens). There are three examples of that pattern. The top two lines expose another pattern. In this pattern, students see how four groups of 9 tens and nine groups of 4 tens are related. To maximize this activity, record lines and related equations in a table or chart (as shown below) and discuss with students the patterns that they observe. Ask them how these patterns can help them find products of tens.

Number	Rod	Total Length	Equation
4 rods	9 cm	36 cm	4 x 9 = 36
4 rods	90 mm	360 mm	4 x 90 = 360

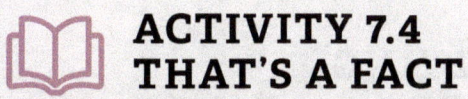

ACTIVITY 7.4
THAT'S A FACT

Some students make the connections with minimal exploration of models, while others need more experience to build a solid foundation. Activities 7.1, 7.2, and 7.3 provide these concrete experiences. This activity provides an opportunity for students to explore number patterns (without the models).

To begin, pose a basic fact and similar problems with factors that are multiples of tens and hundreds, as shown here:

6 × 4 = 24	6 × 40 = 240	6 × 400 = 2,400

Ask students what they notice about the problems. Ask them to describe the similarities and differences between the problems and whether they think other basic facts can be connected to problems like this one (6 × 4). Pose a new fact, having them generate a similar set of problems connected to that fact; for instance, you might pose 4 × 8 = 32. After discussing this second set of problems, determine if another set is needed or if your students are ready to shift to the multiplication charts.

The multiplication charts show the basic facts and related problems with multiples of tens and hundreds. Have your students find the products of the problems you discussed and record them on the charts as shown.

TEACHING TAKEAWAY

To extend an activity like this, write problems on individual cards. Give one to each student. Then, play a group game such as I Have, Who Has (Activity 3.4).

×	1	2	3	4	5	6	7	8	9	10
1										
2										
3										
4						24				
5										
6										
7										
8				32						
9										
10										

×	1	2	3	4	5	6	7	8	9	10
1										
2										
3										
4						24				
5										
6										
7										
8				32						
9										
10										

×	1	2	3	4	5	6	7	8	9	10
10										
20										
30										
40						240				
50										
60										
70										
80				320						
90										
100										

×	1	2	3	4	5	6	7	8	9	10
100										
200										
300										
400						2,400				
500										
600										
700										
800				3,200						
900										
1000										

This resource can be downloaded at **https://qrs.ly/psf6a5o**

Briefly discuss how the equations are represented on the charts, highlighting the similar/different factors and products. Then, pair students. Have one student give a fact, such as 7 × 3 = 21. Their partner then finds the product on the related charts (7 × 30 and 7 × 300). Then have partners change roles. Once groups have completed all of the cells on each chart, have them come back together as a class to confirm accurate products. Close the discussion with an important question for partners to consider and explain: How does knowing a fact help you solve similar problems? Ask them what they might do to solve a problem such as 5 × 50 when all they have is a basic fact chart.

This same activity can be used with division, asking students to notice patterns:

18 ÷ 3 = 6	180 ÷ 3 = 60	180 ÷ 30= 6	1800 ÷ 30= 60

ACTIVITY 7.5
PROMPTS FOR MULTIPLYING
BY TENS AND HUNDREDS

Use the prompts below as opportunities to develop an understanding of and reasoning with the foundation. Have students use representations and tools to justify their thinking, including base-ten models, number lines, number charts, and so on. After students work with the prompt(s), bring the class together to exchange ideas. Remember that these prompts could be useful for collecting evidence of student understanding.

- Ruthie says if you can remember a basic fact, then you will be able to multiply well with tens and hundreds. What do you think she means? Create examples to support your explanation.

- How would you describe the relationship between 8 × 4, 8 × 40, and 8 × 400?

- Miguel says that he knows 4 × 6 = 24 and he can use that to help him solve 60 × 4. Dennis disagrees, saying they aren't the same problem. Who do you agree with (and why)?

- Create or draw a model or diagram showing how 5 × 7, 5 × 70, and 5 × 700 are related.

- Asha says that when you multiply by tens and hundreds, you can simply add zeros. You aren't adding zeros. What is actually happening? Will this pattern always work?

- How would an array of 6 × 3, 6 × 30, and 6 × 300 be similar and different? Use pictures, words, and numbers to explain your thinking.

- Beth wonders if an area of 4 × 70 would be more or less than an area of 7 × 40. What do you think? Explain how you know.

- A product is 2,400. One of the factors is a multiple of 10 or 100. What could the problem have been?

QUALITY PRACTICE FOR MULTIPLYING
BY TENS AND HUNDREDS

Students need enjoyable practice so that they're eager to do more of it. You want them to take the time to become good at something without the pressure of judgment and grading. This isn't to say that students shouldn't be accountable for practice—they should. You want to see that they are engaged, focused, and making progress. You can do that by playing games, exploring with centers, or discussing a routine or graded activity.

ACTIVITY 7.6

Name: "Three Counts"   **Type:** Routine

About the Routine: This routine is built from The Count (Activity 4.6), with some subtle differences. This routine seems somewhat simplistic, but it does provide skip-counting practice while reinforcing relationships of products when multiplying by tens and hundreds. In the directions, you see that you should record the multiplication expressions that go with each number said in the skip count.

A nice modification is for you to record the numbers and then have one partner come up and record the expressions. In this modification, record the skip count horizontally so that many students can come up to the board and record their expressions at once to keep the routine moving.

3	6	9	12	15	18	21	24	27	30	33	36
1×3	2×3	3×3	4×3	5×3	6×3	7×3	8×3	9×3	10×3	11×3	12×3

Materials: This routine does not require any materials.

Directions:

1. Have students work with a partner.

2. Arrange students in a circle with one partner standing in front of the other.

3. Announce a counting interval (e.g., "We'll count by threes") and who will start the count.

4. Have partners whisper to each other what number they think they'll say. For example, the second pair would say 6, the fourth pair would say 12, and so on.

5. When you say go, count around the circle. As students say their count, record the numbers in a column on the board.

6. After the count is complete, have partners identify the multiplication fact that goes with their count (e.g., 3 × 2 = 6).

7. Then, have the partners switch places. Announce the new counting interval (e.g., "We'll count by thirties") and a different starting point.

8. Have partners whisper the new number they think they'll say, then count.

9. As students count, record the numbers. Have them identify the equations and record them, too.

10. Repeat a third time with another count (e.g., "We'll count by three hundreds").

The examples below show the recordings for the Three Counts routine. As students get more comfortable with multiplying by tens and hundreds, have them tell you the equation or write it themselves, as mentioned in the activity introduction.

(Continued)

First Count		Second Count		Third Count	
1 × 3	3	1 × 30	30	1 × 300	300
2 × 3	6	2 × 30	60	2 × 300	600
3 × 3	9	3 × 30	90	3 × 300	900
4 × 3	12	4 × 30	120	4 × 300	1,200
5 × 3	15	5 × 30	150	5 × 300	1,500
6 × 3	18	6 × 30	180	6 × 300	1,800
7 × 3	21	7 × 30	210	7 × 300	2,100
8 × 3	24	8 × 30	240	8 × 300	2,400
9 × 3	27	9 × 30	270	9 × 300	2,700
10 × 3	30	10 × 30	300	10 × 300	3,000
11 × 3	33	11 × 30	330	11 × 300	3,300
12 × 3	36	12 × 30	360	12 × 300	3,600

ACTIVITY 7.7

Name: "Complex Number Strings"   **Type:** Routine

About the Routine: A number string is a list of equations that highlight a pattern within sums and differences, products and sums, and so on. A complex number string is a matrix of equations that presents patterns both vertically and horizontally. See the example below to see how this is presented. They are fairly easy to create. Choose a basic fact and manipulate the first or second factor. The second factor is adjusted in the example below. Notice that the factors in the first column do not change by 1; instead, the first three double (2, 4, 8) then the last two use the previous two rows (4 + 8 = 12, 12 + 8 = 20). You can design these however you like. You can reverse the columns (having 1 × 1 in the upper-right instead of the upper-left). You can also use this routine to practice division patterns.

Materials: A complex number string written on the board or displayed

Directions: 1. Provide a matrix of related expressions with one known product. The first row in this routine is always left blank and found first as an anchor for understanding and as an additional support for the debriefing conversation.

2. Give the class a choice of working down the first column or across the second row (remember that the first row is done together). Students use the known product to work across the rows or down the columns.

3. After students signal that they know the products of a row or column, hold a class discussion about how each known relates to the others. Draw students' attention to how basic facts relate to multiplying multiples of tens, hundreds, and thousands. Keep in mind that students will recognize the pattern of zeros in the products. Be sure to reinforce why this pattern makes sense (e.g., 3 × 40 is three groups of 4 tens).

4. After completing and recording the products for an entire row or column (determined by the class), have students complete the remaining cells.

1 × 1 =	1 × 10 =	1 × 100 =	10 × 10 =
3 × 1 =	3 × 10 =	3 × 300 =	30 × 30 =
3 × 2 =	3 × 20 =	3 × 200 =	30 × 20 =
3 × 4 =	3 × 40 =	3 × 400 =	30 × 40 =
3 × 8 =	3 × 80 =	3 × 800 =	30 × 80 =
3 × 12 =	3 × 120 =	3 × 1,200 =	30 × 120 =
3 × 20 =	3 × 200 =	3 × 2,000 =	30 × 200 =

A complex number string is also perfect for division. The string above could be modified, as shown below:

1 ÷ 1 =	10 ÷ 1 =	100 ÷ 1 =	100 ÷ 10 =
3 ÷ 3 =	30 ÷ 3 =	300 ÷ 3 =	300 ÷ 30 =
6 ÷ 3 =	60 ÷ 3 =	600 ÷ 3 =	600 ÷ 30 =
12 ÷ 3 =	120 ÷ 3 =	1,200 ÷ 3 =	1,200 ÷ 30 =
24 ÷ 3 =	240 ÷ 3 =	2,400 ÷ 3 =	2,400 ÷ 30 =
36 ÷ 3 =	360 ÷ 3 =	3,600 ÷ 3 =	3,600 ÷ 30 =
60 ÷ 3 =	600 ÷ 3 =	6,000 ÷ 3 =	6,000 ÷ 30 =

ACTIVITY 7.8

Name: Six Charts

Type: Game

About the Game: It takes time for students to acquire automaticity with basic facts. Even students with basic fact automaticity may not recognize those facts (e.g., 5 × 7) when one of the factors is a multiple of 10 (5 × 70), 100 (5 ×700), or even decimals (5 × 0.7) and integers (-5 × -70). This game features six related charts connected to basic facts in the upper-lefthand corner. Students roll two factors to find a product to record on one of the charts, working to get a line of products on three of the charts. You can easily adjust the factors and the given products on any of the charts.

Materials: Digit cards (0–9), playing cards (queens = 0, aces = 1; remove tens, kings, and jacks), or a 10-sided die; *Six Charts* game board

Directions:
1. Players take turns rolling two digits. The digits can represent ones, tens, or hundreds. For example, Jake rolls 4 and 6 and can use the 4 as 4, 40, or 400. He can use the 6 in the same way.

2. Players find the product of the digits and how they are viewed. Jake uses his 4 and 6 as 40 and 6 and covers 240.

3. If a product is already covered, the player loses their turn.

4. The first player to cover a line of products on three of the charts wins.

The examples show two options for the game. The left shows a version with factors of ones, tens, and hundreds. The right shows a version with factors of tenths, ones, and tens. You can adjust factors in all sorts of ways to reinforce connections and build student capacity.

> **TEACHING TAKEAWAY**
> You can modify games to provide results to help students see patterns. Each chart on the game board has some prefilled products. Those free spaces are intentionally provided to help students see patterns.

Six Charts

Directions: Players generate two numbers. The numbers can represent ones, tens, or hundreds. The player fills the corresponding product on one of the charts. The first player to get a line of products on three of the charts wins the game.

x	5	6	7	8	9
2	10		14		
3			21		
4		24			
5					45
6					
7	35				

x	500	600	700	800	900
2	1,000		1,400		
3			2,100		
4					3,600
5				4,000	
6	3,000				
7					

x	50	60	70	80	90
2	100		140		
3			210		
4					
5			350		
6					
7	350				

x	5	6	7	8	9
200	1,000		1,400		
300					
400	2,000				
500				4,000	
600					
700					

x	5	6	7	8	9
20	100		140		
30			210		
40			320		
50					450
60		360			
70			490		

x	50	60	70	80	90
20	1,000		1,400		
30			2,100		
40					
50					
60					4,200
70					6,300

Six Charts

Directions: Players generate two numbers. The numbers can represent tenths, ones, or tenths. The player fills the corresponding product on one of the charts. The first player to get a line of products on three of the charts wins the game.

x	5	6	7	8	9
2	10		14		
3			21		
4		24			
5					45
6					
7	35				

x	50	60	70	80	90
2	100		140		
3			210		
4					360
5			400		
6	300				
7					

x	5	6	7	8	9
20	100		140		
30			210		
40					
50			350		
60					
70	350				

x	0.5	0.6	0.7	0.8	0.9
20	1.0		1.4		
30					
40	2.0				
50				4.0	
60					
70					

x	0.5	0.6	0.7	0.8	0.9
2	1.0		1.4		
3			2.1		
4				3.2	
5					4.5
6		3.6			
7			4.9		

x	0.5	0.6	0.7	0.8	0.9
0.2	0.1		0.14		
0.3			0.21		
0.4					
0.5					
0.6					0.42
0.7					0.63

This resource can be downloaded at **https://qrs.ly/psf6a5o**

ACTIVITY 7.9

Name: *Paper Clip Products*　　　　　　　　　　　　　**Type:** *Game*

About the Game: *Paper Clip Products* is a strategy-based game that's easy to learn. It focuses on finding products of multiples of tens while also attending to division. Though only one answer is found with each turn, students subtly practice many more as they think about which move would be best for them. Remember that like all other games, it's a good idea to have students process what they did in the game. After playing, have students write about how the factors and products in this game are related to basic facts.

Materials: Two paper clips, centimeter cubes or something similar for game pieces, one *Paper Clip Products* game board per two players

Directions:　1. The game begins with Player 1 placing a paper clip on a factor in the bottom row. Then, Player 2 places a paper clip on a factor in the left column. Player 1 places their marker on the corresponding product.

　　　　　　　2. It is then Player 2's turn. Player 2 can move each paper clip one space in any direction or they can move one paper clip two spaces in any direction. Player 2 puts their marker on the corresponding product.

　　　　　　　3. Player 1 then moves each paper clip one space or one paper clip two spaces and places their marker on the corresponding product.

　　　　　　　　Note: Players who catch and correct their opponent's error steal the space.

　　　　　　　4. The first player to get five game pieces in a row wins.

The images show how the game is played. Player 1 placed their marker on 300 because it's the product of 5 and 60. Then, Player 2 moved the paper clip in the column of factors one space from 5 to 6 and the other paper clip one space from 60 to 50, covering 300 (6 × 50). Player 1 moved the paper clip in the row two spaces from 50 to 70 to cover 420 (6 × 70). Player 2 then moved the paper clip in the column two spaces to 4, covering 280 (4 × 70). You can modify the order of factors on the outside of the game board for a different playing experience. However, you should keep the products on the game board in the same order so that they are easier to locate during play.

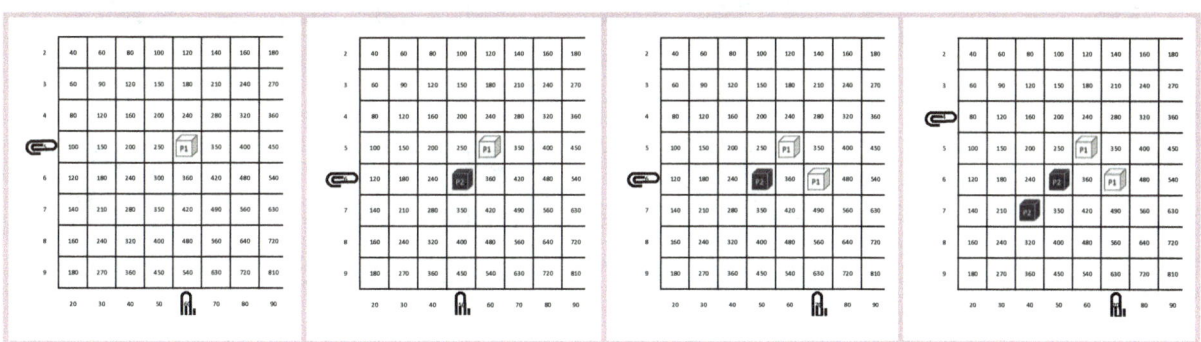

Various game boards for this activity can be downloaded at **https://qrs.ly/psf6a5o**

ACTIVITY 7.10

Name: For Keeps   **Type:** Game

About the Game: *For Keeps* is a fun game of strategy and chance. In this version of the game, students not only practice multiplying by tens and hundreds but they also practice estimating and reasoning. The goal of the game is to get the highest score by adding four products. The player decides which four products to keep with each turn. Note that each player must play eight turns, regardless of how many turns it takes to find four kept products. This is because you want them to practice!

Materials: *For Keeps* game board; digit cards (0–9), playing cards (queens = 0, aces = 1; remove tens, kings, and jacks), or a 10-sided die

Directions:
1. Players take turns generating two digits with a die or cards.

2. Each digit is used to determine a factor. Note that the size of the factor is based on the version of the game students are playing, as shown below.

3. The player finds the product and decides whether they want to keep it as one of their four scores. A player can't change their mind on later turns.

4. After their eight turns, players find the sum of their kept products. The sum is the score.

5. The player with the greatest score wins.

You can create a fun alternative by changing the rules so that the lowest score wins. Of course, there are other adjustments you can also make to practice with different multiples, as shown in the examples. In each version, the player rolled a 3 (first factor) and an 8 (second factor). You can see how the game changes as each digit represents a number of ones, tens, or hundreds.

Round	First Factor (n)	Second Factor (n × 10)	PRODUCT KEEP	PRODUCT DON'T KEEP
	3	80	240	

Round	First Factor (n)	Second Factor (n × 10)	PRODUCT KEEP	PRODUCT DON'T KEEP
	3	800	2,400	

Round	First Factor (n)	Second Factor (n × 10)	PRODUCT KEEP	PRODUCT DON'T KEEP
	30	80	2,400	

As you might suspect, *For Keeps* can be easily modified to feature factors of ones, tens, and hundreds. Students who need practice with decimals could play a version of the game where one factor is a whole number and the other is a tenth or hundredth (e.g., 3 × 0.8 or 3 × 0.08).

For Keeps

Directions: Roll a number for your first factor. Roll a second number to create a factor of ten. Find the product of your two factors. Decide if you want to keep it. You must roll eight times, but you can only keep four products. Add the four products you keep to find your final score. The player with the greatest score wins.

Round	First Factor (n)	Second Factor (n × 10)	PRODUCT KEEP	PRODUCT DON'T KEEP
	3	80	240	
1				
2				
3				
4				
5				
6				
7				
8				
SCORE --------> Sum of the four products I kept				

For Keeps

Directions: Roll a number for your first factor. Roll a second number to create a factor of hundreds. Find the product of your two factors. Decide if you want to keep it. You must roll eight times, but you can only keep four products. Add the four products you keep to find your score. The player with the greatest score wins.

Round	First Factor (n)	Second Factor (n × 100)	PRODUCT KEEP	PRODUCT DON'T KEEP
	3	800	2,400	
1				
2				
3				
4				
5				
6				
7				
8				
SCORE --------> Sum of the four products I kept				

For Keeps

Directions: Roll twice to find the number of tens for your first and second factors. Find the product of your two factors. Decide if you want to keep it. You must roll eight times, but you can only keep four products. Add the four products you keep to find your score. The player with the greatest score wins.

Round	First Factor (n × 10)	Second Factor (n × 10)	PRODUCT KEEP	PRODUCT DON'T KEEP
	30	80	2,400	
1				
2				
3				
4				
5				
6				
7				
8				
SCORE --------> Sum of the four products I kept				

online resources 🖰 This resource can be downloaded at **https://qrs.ly/psf6a5o**

ACTIVITY 7.11

Name: Multiples Bingo   **Type:** Game

About the Game: Bingo is a traditional game that many students find fun. Here, it is used to practice finding multiples of tens or hundreds. Laminate the downloadable bingo board so that students can reuse them with dry erase markers or have them create their own bingo boards to lessen the amount of prep work you need to do. This game is easy to play over and over again by swapping out the multiples. Early play should use single-digit factors (e.g., 7 × 50), but in time, you can work in factors up to 12 (12 × 50), 15 (15 × 50), or even 20 (20 × 50).

Materials: Bingo board, 10-sided dice or digit cards, two-color counters or bingo chips

TEACHING TAKEAWAY

Games such as bingo are good whole-class games. You can also use them for small-group games where five or six students play a game together.

Directions:

1. Note that these directions are for multiples of 50. The game can be played with any multiple of 10 to 100 (e.g., 30, 70, 200, etc.).

2. Optional: Quickly skip count by 50s and record the multiples on the board.

3. Have students write multiples randomly on their game board. They can write a multiple more than once or not at all.

4. When ready, begin play by rolling a number to multiply by 50.

5. Students cover the space on their board that matches the product of that number and 50.

6. Repeat until a player gets a bingo vertically, horizontally, or diagonally. Keep track of how many times a given multiple is called to ensure legitimacy. To do this, you can put tally marks under the multiples you recorded on the board.

The student on the left was playing multiples of 50 (as described in the directions). They recorded multiples of 50 randomly on their game board. Their teacher rolled 7 for the first turn. The class agreed that 7 × 50 was 350. This student covered the lower-center 350 as shown.

Multiples Bingo

350	150	350	50	200
400	250	450	300	500
450	350	250	100	350
250	150	(350)	300	350
200	350	150	400	500

online resources → Blank game boards for this activity can be downloaded at **https://qrs.ly/psf6a5o**

ACTIVITY 7.12

Name: Is the Same, Is the Same   **Type:** Center

About the Center: This center is a practice opportunity wherein students generate problems and rewrite them. The center helps students cement their understanding of multiplying multiples of tens and hundreds by having students decompose a multiple of ten (e.g., 80) into a product of ten (e.g., 8×10). It is easy to reuse, minimizing preparation time. It's perfect for early finishers or for stations and independent practice. A recording sheet is provided, but students could easily create columns in their mathematics journals and record their work there. Note that you might replace *is the same as* with = for students recording in their journal so that they spend more time with mathematics than copying words.

Materials: Digit cards (0–9), playing cards (queens = 0, aces = 1; remove tens, kings, and jacks), or a 10-sided die; Is the Same, Is the Same recording sheet

Directions: 1. Students draw two cards or roll the die twice to generate two numbers and write each in the blank spaces on the Is the Same, Is the Same recording sheet.

2. Students rewrite the problem, decomposing the second factor.

3. Students rewrite the problem after finding the product of the first two factors.

4. Students then write the product.

5. Students repeat this for nine more problems.

(Continued)

The examples show two options for the recording sheet. In both, the multiples of 10 and 100 are the second factor. This is a good place to start. However, you should begin to work in recordings where multiples of 10 and 100 are the first factor; for example, 6 × 80 would be posed as 80 × 6.

A division alternative is shown in the two lower examples. Note that students don't roll digits in that version. Instead, you need to prepare division expressions on index cards for them to pull randomly before recording their work.

_6_ × _8_ 0	is the same as	6 × 8 × 10	is the same as	48 × 10	which is	480
__ × __0	is the same as		is the same as		which is	

_6_ × _8_00	is the same as	6 × 8 × 100	is the same as	48 × 100	which is	4,800
__ × __00	is the same as		is the same as		which is	

My card				
480 ÷ 60	is the same as	480 ÷ 10 ÷ 6	which is	8

My card				
4,800 ÷ 60	is the same as	4,800 ÷ 10 ÷ 6	which is	80

online resources ↖

Recording sheets for this activity that include the above example rows can be downloaded at **https://qrs.ly/psf6a5o**

NOTES

Multiples and Factors

MULTIPLES AND FACTORS OVERVIEW

Multiples and factors are easily confused, so let's start with defining each one.

Multiples are the products for a given number; for example, the multiples of 7 are 0, 7, 14, 21, 28, and so on. Any number that is divisible by 7 is a multiple of 7. *Factors* are those numbers that can divide a number and lead to a whole-number answer. For example, the factors of 30 are 1, 2, 3, 5, 6, 10, 15, and 30. Factors can be multiplied to equal the original number (e.g., 1×30, 5×6, $2 \times 3 \times 5$, and so on). We use multiples and factors for many mathematical topics, in particular for working with multiplication, division, fractions, and ratios. Unfortunately, lessons on multiples and factors can be taught in isolation and by memorizing divisibility rules.

This module is provided because utility with factors and multiples is a necessary ingredient for being fluent in mathematics. Computationally, recognizing multiples and factors is used with multi-digit multiplication and division as well as addition and subtraction of fractions. In time, students will apply this to a range of topics from ratio and proportions to solving equations and factoring polynomials. A strong knowledge of factors and multiples, and skill at finding common factors, is a necessary foundation! Saying multiples by rote does not accomplish the necessary skill set. Students need to have extensive experience with multiples and factors so that when they see numbers such as 18 and 27 in fractions, they recognize them as multiples of 9 and use that information to simplify the fractions, find common denominators, or whatever the task involves. Thus, we need to spend lots of time doing meaningful activities that involve noticing factors and multiples.

Knowing factors and multiples, combined with flexible decomposition of numbers (Module 2), automaticity with multiplication facts, and strengths with multiplying tens and hundreds (Module 7), provides strong support for division. The chart below gives some examples of reasoning with factors, in which you can see good beginnings for strategies to divide using partial quotients.

Is 6 a factor of 72, 186, or 252?	
Rule: Is it divisible by 2 and 3? • 72 is even: 7 + 2 is 9 and 9 is divisible by 3 • 186 is even: 1 + 8 + 6 = 15 • 252 is even: 2 + 5 + 2 = 9	**Reasoning: Can I decompose the number into multiples of 6?** • 72 can be broken into 60 and 12 • 186 can be broken into 180 and 6 • 252 can be broken into 240 and 12
Is 798 a multiple of 7?	
Rule: Double the ones place, subtract it from the other places. If the number is a multiple of 7, then the original number is a multiple of 7. • Double 8 is 16 • 79 – 16 = 63 • 63 ÷ 7 = 9	**Reasoning: Can I decompose the number into multiples of 7?** • 700 is a multiple of 7 • 70 is a multiple of 7 • 28 is a multiple of 7

Is 93 or 378 divisible by 9?	
Rule: Is the sum of the digits divisible by 9? • No. 9 + 3 = 12 • Yes. 3 + 7 + 8 = 18	Reasoning: Can I decompose the number into multiples of 9? • 93 is not because 90 is a multiple and 3 more isn't enough. • 378 can be broken into 360 and 18. Both 360 and 18 are multiples of 9.

What's notable about reasoning through factors and multiples is how much is grounded in having a good beginning with basic facts—automaticity with multiplication facts (and the related division facts) is necessary for extending students' foundations with multiples and factors. This means that it is always worthwhile to continue to provide opportunities to practice facts (and their strategies)!

HOW DO MULTIPLES AND FACTORS CONTRIBUTE TO FLUENCY STRATEGIES?
··

The preceding chart gives you a good idea about how skill with multiples and factors contributes to reasoning with division. Let's explore two examples. On the left, a student solves 648 ÷ 6 by decomposing 648 into 600, 30, and 18; finding partial quotients that are readily divisible by 6; and then combining the partial quotients to get 108. On the right, a student similarly uses partials, first noticing the closest hundred (12 hundreds) that is a multiple of 6. The remaining 119 required writing down partials (60, 30, and 29). Both students are relying on multiples of 6 as ways to navigate these problems.

$648 \div 6$

$600 \div 6 = 100$

$30 \div 6 = 5$

$18 \div 6 = 3$

$100 + 5 + 3 = 108$

$1{,}319 \div 6$ 119

$1{,}200 \div 6 = 200$ 60 30 29

$60 \div 6 = 10$

$30 \div 6 = 5$ 24 [5]

$24 \div 6 = 4$

$200 + 10 + 5 + 4 = 219$

$1{,}319 \div 6 = 219 \text{ r } 5$

The multiples and factors reference page provides an overview, important representations, connections to fluency, and actions to avoid. It can be downloaded and used for reference in planning, teaching, and discussions with colleagues and families.

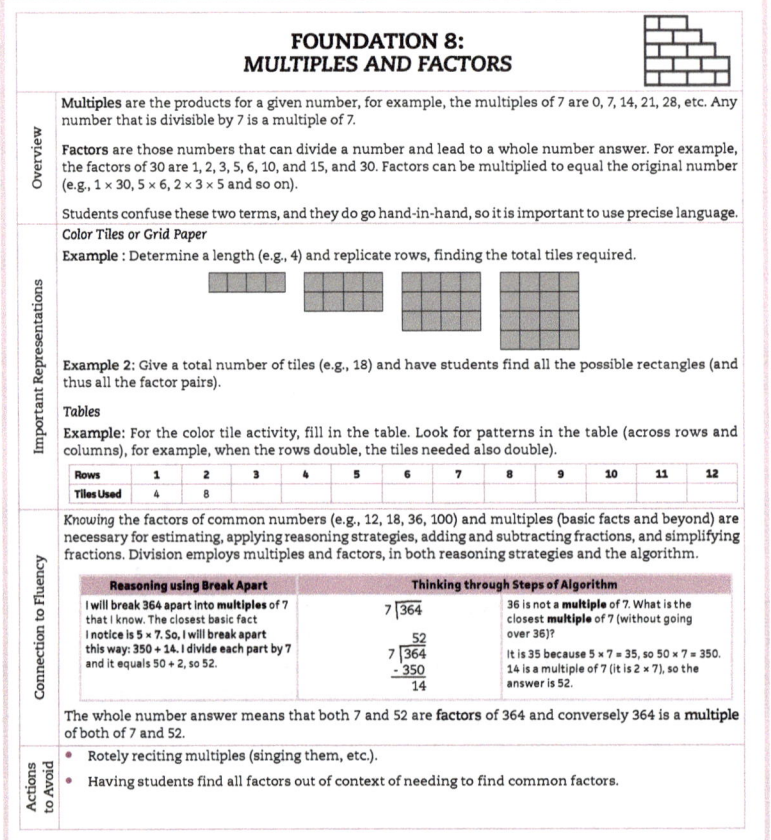

online resources — Available for download at **https://online qrs.ly/psf6a5o**

ASSESSING MULTIPLES AND FACTORS

Assessing multiples and factors can be enacted separately or together. The assessment is more than asking students to say their sevens facts (multiples) but is instead making sure students understand what a multiple is (or what a factor is).

MULTIPLES AND FACTORS LOOK-FORS

It is often apparent if a student is skilled with multiples and factors. You see them make use of their understanding in class as they solve problems, find the volume of a figure, or create equivalent fractions. Their strengths in this arena are typically tied to basic fact fluency. It is unreasonable to expect students to determine any (or every number) to be a factor or multiple of a given number. Instead, the focus is on identifying more obvious factors and multiples (e.g., 84 is a multiple of 4 but not 8, 72 is a multiple of 6 because it is 60 and 12 more, 360 is a multiple of 6 because 6 × 6 = 36 and 360 is 10 × 36). Look-fors include whether a student can

- identify multiples and factors,

- derive a multiple or factor by extending or connecting to a basic multiplication fact,

- determine a multiple or factor through a pattern, and

- explain how they know that a number is a factor/multiple of another without using a rote method.

QUICK ASSESSMENTS FOR MULTIPLES AND FACTORS

The prompts in the table below, along with the routines, games, and centers, are good opportunities to gather evidence of your students' progress. You can do these in small groups or in a one-on-one setting. They should not take more than a few minutes. You have options, but don't feel obligated to use them all. Instead, use a couple and go to a third if the results are inconclusive or if later observations indicate that your student(s) might need additional work with the skill or concept.

Tell the student to describe what a factor is and what a multiple is. Then, ask them to give you examples of each.	Show 36, 51, 24, and 48. Ask the student to tell you which of these is a multiple of 2, 3, and 5. Have them tell you how they know.	Ask what 10×5 is and what 3×5 is. After the correct products are shared, ask what 13×5 is and how they know.
Give students two numbers, such as 15 and 30 or 24 and 36, and ask what is alike and different about these numbers. If they don't focus on multiples and factors, prompt for such responses.	Give students a number (such as 72) and ask them to identify some of its factors. Ask them to explain how they found their factors.	Write down $40 \div 4$ and $24 \div 4$. Ask the student what each quotient is. Then ask the student what they think $64 \div 4$ is and have them explain their thinking.

EXPLICIT INSTRUCTION FOR MULTIPLES AND FACTORS

Decomposing numbers into recognizable multiples of other numbers must be taught to students. You want to anchor this in understanding of what multiplication is and how the distributive property works. After students have an opportunity to look for patterns, be sure that you point them out. You can teach the idea in a lesson or two, but know that your students will need lots of practice and repetition. Keep in mind that caregivers might not be familiar with the idea of decomposing numbers for this purpose. You might give them a few simple examples (e.g., 72 is a factor of 6 because it can be broken into 60 and 12) so that they understand and can support practice at home.

NOTES

_____

_____

_____

_____

_____

_____

ACTIVITY 8.1
BUILDING MULTIPLES

Ideas of factors and multiples begin with a conceptual understanding of multiplication. This activity is a hands-on experience that may be familiar. In it, students build area models to identify multiples of a given factor and record the equations. Students then create a table to show the factors (3 and the number of rows) and corresponding multiples. In the example, partners have built five multiples of 3.

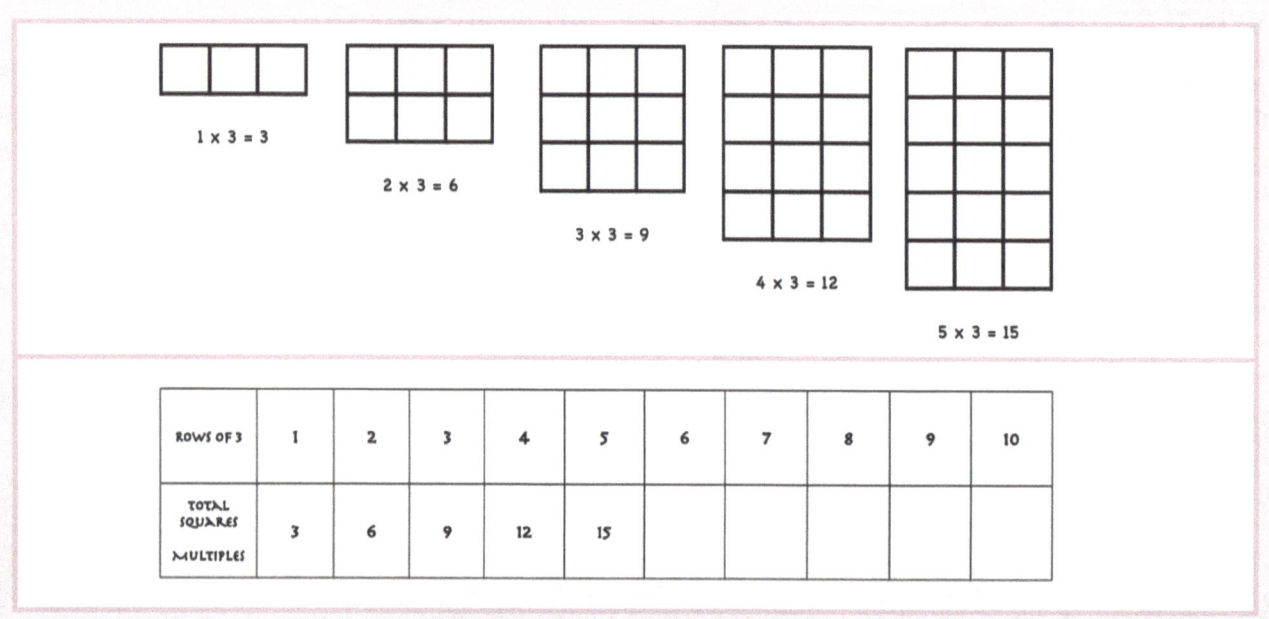

ROWS OF 3	1	2	3	4	5	6	7	8	9	10
TOTAL SQUARES MULTIPLES	3	6	9	12	15					

The rectangles built in the example are made with color tiles. It could be a challenge to gather enough color tiles for ten groups in your class. To offset this challenge, students can use graph paper to make their models. In the following example, students have shaded their rectangles. Alternatively, they could cut out their rectangles. Either of these options can be especially useful as they begin to work with larger multiples or when they start to break apart numbers, as described in the next activity.

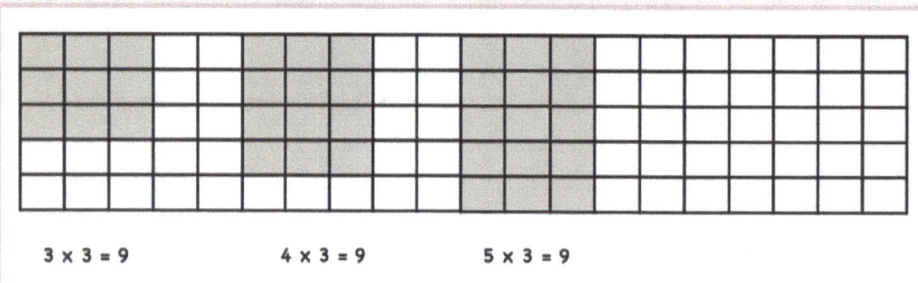

ACTIVITY 8.2
BREAKS IT, MAKES IT

Divisibility rules are one way for determining whether a number is a factor or multiple of another number. While those rules do work, they don't call on nor do they reinforce students' number sense and flexibility. This lesson is an introduction to a simple premise that is overlooked by many who have experienced mathematics in a closed, rote, and procedural way. This important idea is that you can determine if a number is a multiple of another number by breaking it into benchmark, recognizable multiples of that number.

For example, how do you know that 72 is a multiple of 6? Do you look to see that the sum of its digits is 9 (divisible by 3) and that it is an even number? Or do you know that it is composed of 60 and 12 and both are multiples of 6? Does it still work if you break 72 into 30, 30, 6, and 6? Yes! It does. There are other ways to decompose it as well. (Hint: Think of how many hours are in three days.) The simple point is that teaching students to decompose a number into multiples promotes their agency, flexibility, and number sense.

Begin this activity by identifying and recording multiples of 4 to 40. Then, pose the question to students, "If 40 is a multiple of 4, do you think 80 is a multiple of 4?" Have students talk with partners, form an argument, and share their ideas. Discuss it with the class. Make it clear during the discussion that 80 is a multiple of 4 because you can break it into 40 and 40. From there, ask if there are other numbers that can be identified as multiples of 4 by breaking them into parts.

Pose a table for students to explore. Encourage them to come up with more than one decomposition when possible. A second option for 80 is offered in the table. Some numbers might only have one convenient combination. For example, 80 could be broken into 64 + 16 or 24 + 24 + 32, but those combinations aren't necessarily convenient. The table shows possible decompositions in italics. The chart should be provided to partners with blank cells.

IS ____ A MULTIPLE OF 4?	YOU CAN BREAK IT INTO	YOU CAN BREAK IT INTO
80	*40 + 40; 40 is a multiple of 4*	*20 + 20 + 20 + 20; 20 is a multiple of 4*
56	*40 + 16 are both multiples of 4*	*20 + 20 + 16; 20 and 16 are multiples of 4*
72	*20 + 20 + 20 + 12; 20 and 12 are multiples of 4*	*80 – 8; 80 is a multiple of 4 and then you take a multiple of 4 away*
96	*80 + 16; 80 and 16 are multiples of 4*	*100 – 4; 100 is a multiple of 4 and then you take a multiple of 4 away*
116	*100 + 16; 100 and 16 are multiples of 4*	*120 – 4; 120 is a multiple of 4 and then you take a multiple of 4 away*
168	*160 + 8; 160 and 8 are multiples of 4*	*80 + 80 + 8; 80 and 8 are multiples of 4*

This activity easily extends to other numbers. After working with multiples of 4, move to multiples of 3, 6, and 8. For example, 81 is a multiple of 3 because it can be decomposed into 60 + 21 or 30 + 30 + 21. Keep in mind that you want to do this work over a period of time. Be patient; allow time for students' fact recall to develop. Along the way, provide fact charts and other tools to support them. Also, think about if and how you would incorporate other multiples, such as 2, 5, 7, 9, or 10.

If needed, you can introduce the concept of breaking multiples apart by building models of multiples, described in Activity 8.1. The image shows how a student could think about 72 breaking it into 30, 30, and 12.

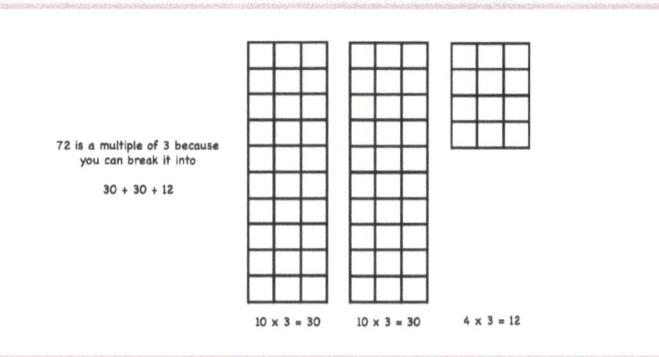

72 is a multiple of 3 because you can break it into

30 + 30 + 12

10 × 3 = 30   10 × 3 = 30   4 × 3 = 12

ACTIVITY 8.3
MULTIPLES VENN GALLERY WALK

This task is more about engagement, discussion, and self-assessment rather than instruction. Put students into groups of three or four. Assign two factors to the group. Have them create a chart of the first twenty multiples of each factor on a piece of paper and turn it over. Note that multiplication charts and calculators should not be used for this activity. Once each group has their multiples recorded, have them create a large Venn diagram with their two factors for the rings.

Each group posts their Venn diagram. Groups move from chart to chart, adding three or four new multiples to the chart. They can add multiples beyond the first twenty. After each poster is visited, groups use their original list of factors to check the multiples recorded on their poster and determine if other multiples are correct. Groups then report the accuracy of their poster to the class. They should share multiples that were missing or incorrect. There may be moments when a group isn't sure about the accuracy of a multiple. When this happens, charge the whole class with determining whether it is or isn't accurate, along with providing an explanation.

6	4
6, 12, 18, 24	4, 8, 12, 16,
30, 36, 42, 48	20, 24, 28, 32
54, 60, 66, 72,	36, 40, 44, 48
78, 84, 90, 96,	52, 56, 60, 64,
102, 108, 114, 120	68, 72, 76, 80

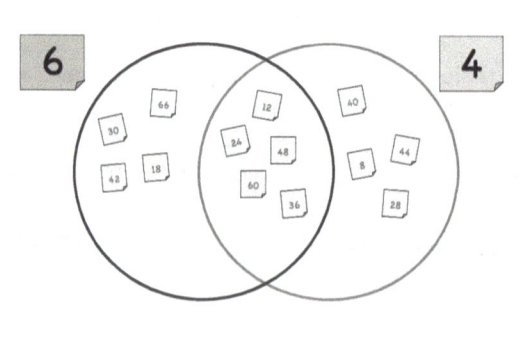

ACTIVITY 8.4
BUILDING TABLES OF SPECIAL MULTIPLES

Certain factors and their multiples are prevalent in mathematics and everyday life. Some of those factors include 12, 15, 24, 25, 50, and 75. Your curriculum probably doesn't call for students to have experience with or to use these numbers as factors. But assuming that your students' understanding and automaticity of factors and multiples is developing well, it is a good idea to at least informally explore these factors and their multiples. Be up-front with your students; tell them why these numbers appear so often and share why working with them will help them in their lives and as they move through later classes. If possible, you want your students to know the first three or four multiples of these numbers. In time, that will merge with their number sense and decomposition skill for powerful manipulation, recognition, and computation.

12	Knowing multiples of 12 means one then knows multiples of each of these: 24, 36, 48, and 60. From there, they can derive 84 (60 + 24), 96 (60 + 36), or 156 (120 + 36) as multiples of 12.
15	Knowing multiples of 15 is also very useful, with the most obvious application to time (15 minutes). It is worthwhile for students to develop automaticity with multiples of 15, at least through 15 × 4.
24	Knowing multiples of 24 is very useful. The most obvious connection here is to 24 hours in a day. It is useful for students to develop automaticity with multiples of 24, at least through 24 × 4.
25	Knowing multiples of 25 is useful because it is one-fourth of a hundred, used in money, and is a common fraction and percent. It is useful to first develop automaticity with the first four multiples of 25, and then note the pattern that repeats with every four numbers, noticing that eight of 25 is 200, twelve of 25 is 300, and so on.

An easy instructional activity is to have students create tables of these multiples with partners or small groups. Examples for multiples of 12 and 15 are shared in the following examples. Bring the class together to share their results and discuss the patterns they notice in the tables. Use their work and insights to create anchor charts of these multiples. Then, practice with these using routines such as The Count (Activity 4.6) or The Stand (Activity 1.7) as well as games for practicing with multiples.

Multiples of 12 Table

	n × 1	n × 2	n × 3	n × 4	n × 5	n × 6	n × 7	n × 8	n × 9	n × 10
3	3	6	9	12	15	18	21	24	27	30
6	6	12	18	24	30	36	42	48	54	60
12	12	24	36	48	60	72	84	96	108	120

Multiples of 15 Table

	n × 1	n × 2	n × 3	n × 4	n × 5	n × 6	n × 7	n × 8	n × 9	n × 10
5	5	10	15	20	25	30	35	40	45	50
10	10	20	30	40	50	60	70	80	90	100
15	15	30	45	60	75	90	105	120	135	150

ACTIVITY 8.5
PROMPTS FOR MULTIPLES AND FACTORS

Use the prompts below as opportunities to develop understanding of and reasoning with the strategy. Have students use representations and tools to justify their thinking, including base-ten models, number lines, number charts, and so on. After students work with the prompt(s), bring the class together to exchange ideas. Remember that these could be useful for collecting evidence of student understanding.

- Create a pattern of multiples. Describe your pattern and prove that you know the multiples are correct.

- Xander knows that 72 is a multiple of 6 because you can break it into 60 and 12, and both of those numbers are multiples of 6. Do you agree with Xander's thinking? Create an example to prove your thinking.

- Find three different numbers that have a common multiple. Two of the numbers should be two-digit numbers.

- You know that 40 is a multiple of 4. How can that help you know that 84, 128, 164, or 248 are multiples of 4?

- Rory is finding factors. He knows that 2 is a factor of 156. He wonders if 3, 4, and 6 are also factors. Do you think they are factors of 156? Show how you know.

- Choose a number between 1 and 9. Skip count by that number twenty times and record each number. What pattern do you notice in the numbers?

- Gina thinks 39 is prime because it doesn't have any factors other than 1 and itself. State why you agree or disagree with Gina.

- Predict how many prime numbers you think there are between 1 and 100. Find them all and compare the number with your prediction. Note: You can provide a hundred chart to students to support and organize their thinking.

QUALITY PRACTICE FOR MULTIPLES AND FACTORS

Practicing with accuracy is critical for fortifying students' fluency foundations. Accuracy can be challenging in situations of finding combinations of 100 (Module 3), flexible decomposition (Module 2), and even skip counting (Module 4). It's clearly a challenge to find multiples and factors. To support accuracy, provide number and operation charts. Encourage—even require—students to use calculators. You can have them make their first attempts without a calculator and then check their work. In game settings, players can check the accuracy of their opponent and possibly steal a turn when they find inaccuracies. Keep in mind that using a calculator is not cheating but it is yet another dose of practice.

ACTIVITY 8.6

Name: "The Stand (Multiples Version)"

Type: Routine

About the Routine: The Stand is a whole-class routine or game to practice factors or multiples. Students who make a number that is a multiple of a given factor "survive" the round. But those who don't may be better positioned to win in the long run.

Materials: One deck of digit cards (0–9) or playing cards (queens = 0, aces = 1; remove tens, kings, and jacks) for each student

Directions:

1. The leader (the teacher or a selected student) asks all students to stand and announces a factor (e.g., 4).

2. Each student deals themselves three cards and then attempts to make a multiple of the called number (4). They can use two of the cards or all three. Note that single-digit multiples are not allowed (e.g., 4 or 8 cannot be used as solutions in this example).

3. Using a random method, the leader calls on five or so students.

   - If the student who is called on did not make a multiple of 4, they sit down (temporarily).

   - If the student did form a multiple of 4, they remain standing.

4. All students discard their three cards.

5. The leader again identifies a factor (e.g., 5) and students (both sitting and standing) draw three cards and try to form a multiple (of 5).

6. Students are called on randomly again. The stakes above for standing players remain. But there is one new twist: If a sitting student is selected and they have a multiple of the called factor (e.g., 5), they stand back up and pick three students (or randomly draw three names) to sit down.

7. Repeat with new factors and/or new cards until only one student is left standing. That student is the winner of "The Stand."

Note that when a multiple is shared, you can have the class quickly chat with a partner to confirm that they agree that it is a multiple. For example, a student might say that 96 is a multiple of 6. Students can quickly chat to determine whether it is. You can record proof on the board. For 96, someone might say that 96 is composed of 30, 30, 30, and 6. Each of those is a multiple of 6, so 96 is as well. Someone else might say that 96 is composed of 60 and 36, using the same logic. Of course, a student might cite a divisibility rule, which is also a viable option. Keep in mind that you can pick and choose the multiples you discuss (you don't have to discuss each one).

ACTIVITY 8.7

Name: "Knockout"　　　　　　　　　　　　　　　　**Type:** Routine

About the Routine: Knockout is a playful, game-like routine that pits your students against you. It is perfect for injecting some excitement and energy into your mathematics class. The goal of the activity is for you to knock a certain number of your students out of the game in one, two, or three rounds. Students record multiples of a given number and you try to catch students writing incorrect multiples or duplicates. As you play, it's a good idea to briefly confirm that a number shared by a student is in fact a multiple and how they know that it is. For example, if a student says 96 is a multiple of 4, you would ask how they know or for others in the class to confirm. You can increase motivation by offering a small class prize (e.g., no homework) when they beat you a certain number of times.

Materials: Whiteboards or scrap paper

Directions:　1. Have all your students stand.

2. You announce your knockout goal. For example, "I can knock out five students in three rounds."

3. To begin a round, announce a number (e.g., 4).

4. Have students write down a multiple of that number without looking at any classmates' ideas.

5. Call on a student to share their number. If they share a correct multiple, they stay standing. If not, they are knocked out (i.e., they sit down). This is your first try for the round.

6. Now call on another student; this counts as your second try of the round. If the player has the same multiple as another student you have called on or if their number isn't a multiple, they are knocked out (i.e., they sit down).

7. If a student is knocked out on your try, you get to go again (i.e., call on another student). If the student isn't knocked out, you're on to your next try. You get five tries in a round.

8. When students are knocked out, they still get to play by consulting with a standing student to advise about a multiple the standing student might write down.

You will find that your students will come back to the same numbers again and again. For instance, students might use basic facts solely for their multiples. When this happens, add conditions to the game. For instance, you can have them find multiples of 4 that are greater than 60 but less than 200 (or something similar). This adjustment will require them to think beyond basic facts.

Keep in mind that you can change this game quite easily to take advantage of the amount of time you have set for your routine. For a longer version, try to play three rounds. For a quick version, play one round. To include an element of chance, use a spinner or random generator to determine the number of students you have to knock out in a game.

ACTIVITY 8.8

Name: Multiple Cover Up   **Type:** Game

About the Game: *Multiple Cover Up* is an opportunity for students to practice recalling multiples. Players strategically choose multiples to cover on the game board with the goal of making as many three-in-a-rows as possible before all the cards are used. The game board includes prime numbers (which are marked as free spaces). An alternate version of the game board is provided that does not mark the prime numbers. In that version, students have to be on the lookout for primes and avoid them because they can block a three-in-a-row. After playing, ask, "Which multiples are easiest for you to cover and why?" or "How does finding multiples help you with other things you do in mathematics?"

Materials: *Multiple Cover Up* factor cards, *Multiple Cover Up* game board, centimeter cubes for game pieces

Directions: 1. The goal of this game is to get more three-in-a-rows than your opponent. The prime numbers are free spaces. Either player can use these to make three in a row. For example, a player with a piece on 28 and 30 has three in a row using 28, 29 (free), and 30.

2. Players take turns pulling a *Multiple Cover Up* factor card.

3. On their turn, a player covers any two available multiples of the factor they pull. If there are no spaces available, the player loses their turn.

4. Play until all of the *Multiple Cover Up* factor cards are used.

5. Players count all of their three-in-a-rows. The player with the most three-in-a-rows wins.

The image shows the start of a game with two players. Player 1 (gray pieces) picks a factor of 6 and covers 60 and 66. Player 2 (white pieces) picks a factor card of 12 and covers 24 and 48. The game continues until a player has three in a row. Though the game might seem simple, some numbers are strategic to cover for certain factors.

Multiple Cover Up

Directions: Flip a card. Cover two multiples of the number on the card. When all of the cards have been used, the player with the most three in a rows wins. A free space can be used by either player.

11 Free	12	13 Free	14	15	16	17 Free	18	19 Free	20
21	22	23 Free		25	26	27	28	29 Free	30
31 Free	32	33	34	35	36	37 Free	38	39	40
41 Free	42	43 Free	44	45	46	47 Free		49	50
51	52	53 Free	54	55	56	57 Free	58	59	
61 Free	62	63	64	65		67 Free	68	69	70
71 Free	72	73 Free	74	75	76	77	78	79 Free	80
81	82	83	84	85	86	87	88	89 Free	90
91	92	93	94	95	96	97 Free	98	99	100
101 Free	102	103 Free	104	105	106	107 Free	108	109 Free	110
111	112	113 Free	114	115	116	117	118	119 Free	120

Resources for this activity can be downloaded at **https://qrs.ly/psf6a5o**

ACTIVITY 8.9

Name: Just Three Factors

Type: Game

About the Game: This game is an alternative to *Multiple Cover Up*. It helps students get comfortable with thinking about how a number can be a multiple of many different numbers. Similar to other games, it is good for helping students recognize factors and build toward quicker recall of factors and multiples. But as you know, student accuracy can be problematic. So, in this game, players check their opponent's accuracy with a calculator, giving both players a small dose of practice and extra motivation to be accurate.

Materials: One deck of digit cards (0–9) or playing cards (queens = 0, aces = 1; remove tens, kings, and jacks), *Just Three Factors* game board, and centimeter cubes (or something similar) for game pieces

Directions:

1. Both players use the same game board.

2. Players take turns dealing three cards.

3. Players can rearrange the digits on the cards to make any two- or three-digit number.

4. The player determines what their number is a multiple of. Note that if a player makes 72, they don't have to name all of its factors. They can simply say it's a multiple of 6 and explain how they know this. For example, they could say 6 × 12 is 72 or they could say that 72 can be broken into 60 and 12. Both 60 and 12 are multiples of 6.

5. The player covers one factor on the game board.

6. After a player shares their thinking, their opponent uses a calculator to confirm that the number made is a multiple of the number covered. If inaccurate, the opponent steals the space and play continues.

7. The first player to cover three spaces in a row wins.

TEACHING TAKEAWAY

Modify games so that players can use calculators to check their opponent's accuracy. Allow players to steal turns or spaces when their opponent isn't accurate.

In the example, Player 1 (white cubes) was dealt 3, 5, and 7. They made 75 and said it was a multiple of 5 because it had 5 ones. They placed their piece on a 5. That player could have made 375, 35, or 735 as well. Player 2 (gray cubes) was dealt 6, 1, and 8. They said 168 was a multiple of 8 and covered an 8. Note that multiples run diagonally on the board, so they are easy to find. You can alter the directions by allowing students to make more than one multiple with the three cards dealt. With this modification, you should set a goal of five in a row or more than one set of three in a row.

Just Three Factors

Directions: Flip three cards. Use two or all three to make a multiple of a number on the board. Cover the factor on the game board. Use a calculator to check your opponent's thinking. If they're wrong, you can steal the space. The first player to get three in a row wins.

2	5	3	8	10	12	7	25
4	2	5	3	8	10	12	7
6	4	2	P1	3	P2	10	12
8	6	4	2	5	3	8	10
10	8	6	4	2	5	3	8
12	10	8	6	4	2	5	3
9	12	10	8	6	4	2	5
25	9	12	10	8	6	4	2

Player 1

Player 2

online resources — This resource can be downloaded at **https://qrs.ly/psf6a5o**

ACTIVITY 8.10

Name: The Connects   **Type:** Game

About the Game: This game is similar to the classic game Boggle. First, a number is called. Then, students try to find multiples of that number using digits on their game board. Their digits must be adjacent, similar to spelling a word in Boggle (see the example). You can download the template and have students make their own game board by filling in each blank space with a digit. Alternatively, you can download a template with four game boards and fill in the digits yourself before copying. This is a good option because all students can play on the same board with the same digits. After playing, you can hold class discussions about the multiples they found.

Materials: One game board per student

Directions: 1. Each student has a 5 × 5 grid with a digit in each number.

2. Call out a number that will be a factor.

3. Give students time to find multiples on their game board.

4. Have students record the multiples they find and have them prove that they know it is a multiple.

5. After time is up, have students count the number of multiples they find. Students earn one point per two-digit multiple and two points per three-digit multiple.

6. After playing and scoring, have students share some of the multiples they found. Hold a brief discussion about how they know the number is a multiple.

In the example, students were to look for multiples of 6. This player found 12, knowing that 6 × 2 = 12. They identified 66 diagonally from left to right. They found 48 as shown on the right side. They also found a three-digit multiple (150). Notice how they recorded their proof: They decomposed 6 into 4 and 2, multiplying each by 25. This same thinking could be applied to 72 (not found), thinking of it as 6 × 10 + 6 × 2 or as 6 × 12.

6 x 2 = 12

6 x 11 = 66

6 x 8 = 48

6 x 25 because 4 x 25 = 100 and 2 x 25 = 50 and 100 + 50 = 150

2	3	4	7	9
8	1	2	4	7
3	6	4	8	3
6	2	5	4	3
9	0	1	5	8

 Blank game boards can be downloaded at **https://qrs.ly/psf6a5o**

ACTIVITY 8.11

Name: Five Card Multiples **Type:** Game

About the Game: *Five Card Multiples* is a simple, fun game for practicing multiples. It works well for two to four players. It can be used as a quick game break or students can play multiple rounds for an extended experience. An advantage with this game is that students practice multiples when it's both their turn and their opponent's. As with other games, you should have students check their opponent's accuracy with calculators or focused multiple tables as shown here.

	×1	×2	×3	×4	×5	×6	×7	×8	×9	×10
3	3	6	9	12	15	18	21	24	27	30
	×11 ×10 + ×1	×12 ×10 + ×2	×13 ×10 + ×3	×14 ×10 + ×4	×15 ×10 + ×5	×16 ×10 + ×6	×17 ×10 + ×7	×18 ×10 + ×8	×19 ×10 + ×9	×20 ×10 + ×10
3	33 30 + 3	36 30 + 6	39 30 + 9	42 30 + 12	45 30 + 15	48 30 + 16	51 30 + 21	54 30 + 24	57 30 + 27	60 30 + 30
	×21 ×20 + ×1	×22 ×20 + ×2	×23 ×20 + ×3	×24 ×20 + ×4	×25 ×20 + ×5	×26 ×20 + ×6	×27 ×20 + ×7	×28 ×20 + ×8	×29 ×20 + ×9	×30 ×20 + ×10
3	63 60 + 3	66 60 + 6	69 60 + 9	72 60 + 12	75 60 + 15	78 60 + 16	81 60 + 21	84 60 + 24	87 60 + 27	90 60 + 30

	×1	×2	×3	×4	×5	×6	×7	×8	×9	×10	×11	×12	×13	×14	×15	×16	×17	×18	×19	×20
3	3	6	9	12	15	18	21	24	27	30	33	36	39	42	45	48	51	54	57	60
6	6	12	18	24	30	36	42	48	54	60	66	72	78	84	90	96	102	108	114	120
9	9	18	27	36	45	54	63	72	81	90	99	108	117	126	135	144	153	162	171	180

	×1	×2	×3	×4	×5	×6	×7	×8	×9	×10	×11	×12	×13	×14	×15	×16	×17	×18	×19	×20
2	2	4	6	8	10	12	14	16	18	20	22	24	26	28	30	32	34	36	38	40
4	4	8	12	16	20	24	28	32	36	40	44	48	52	56	60	64	68	72	76	80
8	8	16	24	32	40	48	56	64	72	80	88	96	104	112	120	128	136	144	152	160

Materials: Playing cards (aces = 1, face cards are wild)

Directions:

1. Each player is dealt five cards.

2. The rest of the cards are placed in a deck face down.

3. The top card is flipped over.

4. Player 1 uses one or more cards to make a multiple of the flipped card. They discard the cards they use to make that multiple. Their turn ends and a new card is flipped for Player 2 to make a multiple of.

5. If Player 1 can't make a multiple on their turn, they take a card from the deck and the flipped card goes to the next player, who tries to make a multiple. If Player 2 can make a multiple, they discard the cards used and the next player gets to go.

6. If Player 2 can't make a multiple, they take a card and play moves to the next player trying to make multiple.

7. The game continues until a player discards all of their cards.

TEACHING TAKEAWAY

Consider having students make multiple tables during independent work to use when they play games.

(Continued)

The example below shows play with three players. The first card flipped was the 4 of spades. Player 1 has to make a multiple of 4. Not recognizing that 36 is a multiple of 4, Player 1 says they don't have a multiple. They add a card to their hand and play goes to Player 2. Player 2 has many options because they have a face card that is a wild card (played as any digit). Player 2 could discard their 7 and 2 (because 72 is a multiple of 4) or they could discard their 5 and 2 because 52 is a multiple of 4; they could also use their wild card as any digit and combine it with other numbers to make a multiple. For example, they could use their ace (1), two, and wild card (played as a 0) because 120 is a multiple of 4. However, keeping a wild card has its advantages for later turns. After making their decision, a new card is flipped and Player 3 takes a turn to make a multiple. Player 1 will go next and so on until someone discards all of their cards.

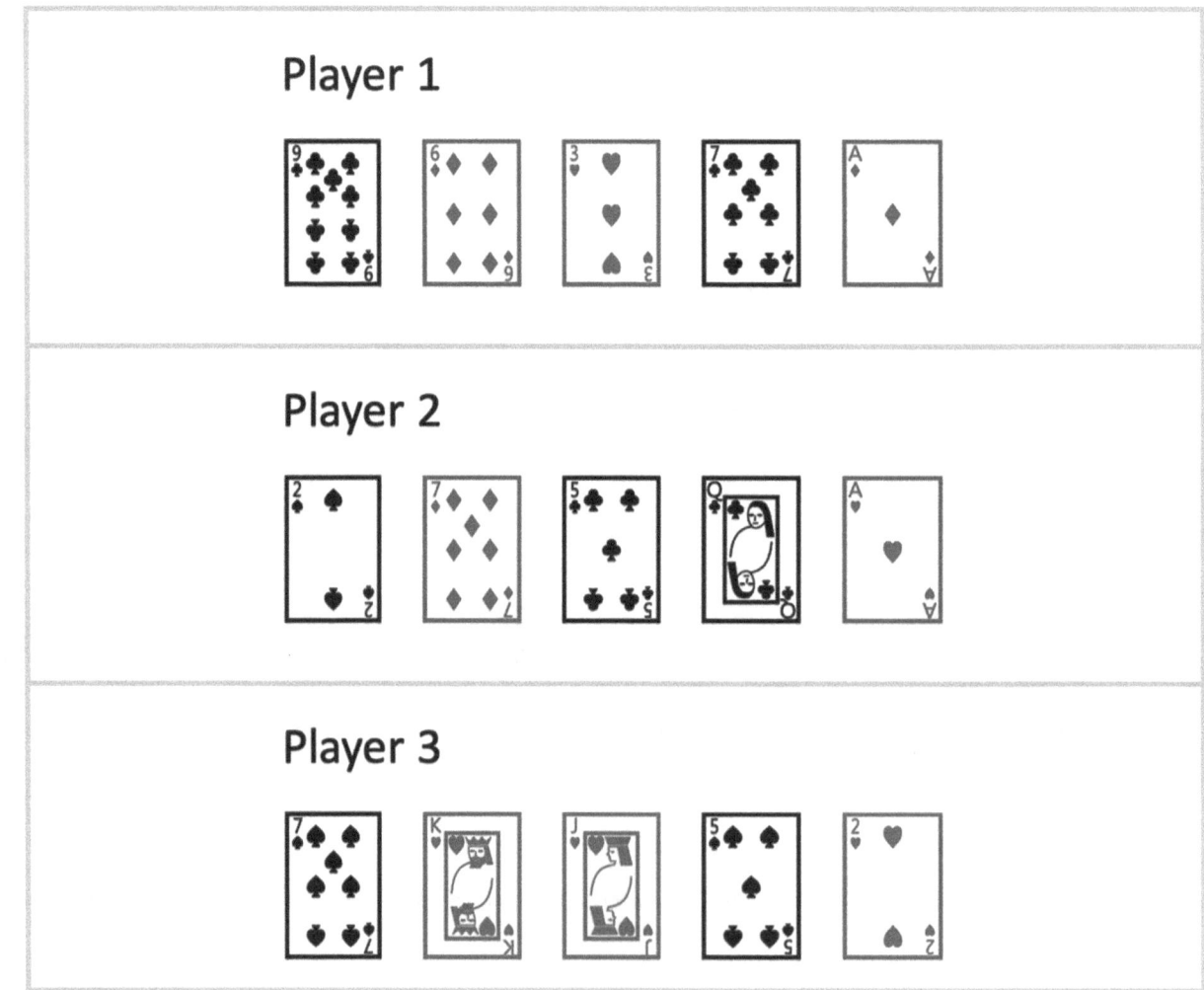

ACTIVITY 8.12

Name: Dropping Primes

Type: Game

About the Game: You can practice factors and multiples indirectly by playing games with prime numbers. In this game, players try to make primes to get rid of the cards in their hand. But players have to be careful of the prime numbers they discard. If their opponent determines that the number isn't prime, they have to take a penalty card—making it harder to clear their hand. You can have opponents check for accuracy with calculators to keep the game moving. You can add a fun twist that if an opponent can determine mentally that the number isn't prime, the penalty is two cards!

Materials: Playing cards (aces = 1, face cards are wild, tens are removed) or a set of *Dropping Primes* cards

> **TEACHING TAKEAWAY**
> Find ways to encourage students to do mathematics mentally when they can. Awarding bonus turns or increasing penalties are ways to do this.

Directions:

1. Each player is dealt three cards.

2. Players take turns trying to make a prime number with two or all three cards in their hand.

3. At the end of each turn, players discard one card.

4. If a player makes a prime number, they discard those cards. Note that a player making a two-digit prime number would discard their third card to clear their hand. Making a three-digit prime automatically clears the hand.

5. If the player can't make a prime number, they take a new card from the deck and discard one of the cards in their hand, holding no more than three cards on any turn unless they make an error.

6. When a player plays a prime number, their opponent checks to see if they are correct. If the number they make isn't a prime, they take a penalty card and will have four cards in their hand for the rest of the game.

7. The first player to discard all of their cards wins, but all players must have equal turns (i.e., if Player 1 goes out first, Player 2 has a chance to go out on their next turn).

First Turn:	
Player 1 is dealt a 4, 2, and 8. Player 1 cannot make a prime number out of these cards. Player 1 gets a king (wild; shown below) and discards the 2 to end their turn.	4♥ 2♣ 8♣
Second Turn: Player 1 uses the king (wild) as a 7. Player 1 says 87 is a prime number. Player 2 disagrees and checks. Player 2 is correct, so Player 1 gets a penalty card (3) and doesn't get to discard to end their turn.	4♥ *Discarded* 8♣ K♠

(Continued)

(Continued)

Third Turn: Player 1 starts this turn with a 4, 8, king, and 3. The 3 is a penalty from last turn.	4♥ 8♣ K 3♦
Third Turn (continued): Player 1 rearranges their cards to make 43 and 83 (the wild card is used as a 3 on this turn instead of a 7).	4♥ 3♦ 8♣ K♥
Player 1 Wins: Player 1 says that 43 is a prime. They are correct and the cards are discarded. Then, they say that 83 is also prime and discards the two cards to clear their hand.	4♥ *Discarded* 3♦ *Discarded* 8♣ *Discarded* K♥ *Discarded*

Optional playing cards can be downloaded at **https://qrs.ly/psf6a5o**

NOTES

Doubling and Halving

DOUBLING AND HALVING OVERVIEW

Do you give it much thought when you double a number? What about when you find half of a number? Doubling and halving are skills we use often, and thus, they are an important foundation.

Doubling starts by learning doubles with the basic facts (first 6 + 6 and later 2 × 6). As students become automatic with these fact sets, they can use the double to derive other facts:

6 + 8: "I know 6 + 6 is 12, so 6 + 8 is 2 more—14" (Near Doubles strategy)

4 × 7: "I know two groups of 7 (2 × 7) is 14, so four groups is twice as much—28." (Doubling strategy)

Doubling grows from basic facts to multi-digit whole numbers. Some numbers, such as 30, 53, and 122, are easy to double; others, such as 48 and 355, are not as easy. Finding half of a number is the same as dividing it by 2. Similar to doubling, some numbers (e.g., 16, 80, and 166) are easier to halve, while others (e.g., 45 and 136) are not as easy. When it is not obvious what the double or half is, decomposing can be used to find the double or half.

Students show these different approaches below to doubling 48 and halving 136.

Double 48 is 96 40 + 40 = 80 8 + 8 = 16	Double 48 is 96 50×2 = 100 100 − 4 = 96 2×2 = 4
136 half is 68 100 → 50 30 → 15 6 → 3 50 + 15 + 3 = 68	120 + 16 = 136 60 + 8 = 68 } half

This module is about doubling or halving numbers. It is not about the strategy of halving one factor and doubling the other, though it obviously builds the foundation to eventually use that strategy. Doubling and halving aren't featured much in mathematics curriculum. Doubles and halves are a fact set—or "fact family"—for addition (4 + 4), subtraction (8 − 4), multiplication (4 × 2), and division (8 ÷ 2). Beyond that, there may be a class discussion about doubling 25 or halving 100 here or there—usually in the context of coins and money—but there is little to no instructional emphasis on doubles and halves anywhere else in elementary curriculum standards.

As mentioned, some numbers are easy to double as students leverage patterns and relationships between 4 and 40 in order to find 8 or 80. But doubling 48 or halving 136 can be a bit of a challenge. Through experience and refined number sense, you may have been able to figure out ways to double and halve. Instead of relying on discovery and chance, your students can learn and practice these skills explicitly through activities in this module. It will serve them well in their everyday lives and their future mathematics classes as they find common denominators, reason about proportionality, estimate, and compute.

HOW DO DOUBLING AND HALVING CONTRIBUTE TO FLUENCY STRATEGIES?

Doubling and halving is a significant fluency strategy (Bay-Williams & SanGiovanni, 2021). To find the product of 28×5 or $6 \times 3\frac{1}{2}$, you can double one factor and halve the other to create a simpler, more efficient problem. Doubling and halving only works with multiplication. Students use the strategy to find those products. On the left, the student halves 28 and doubles 5, creating 14×10. On the right, the student halves 6 and doubles $3\frac{1}{2}$.

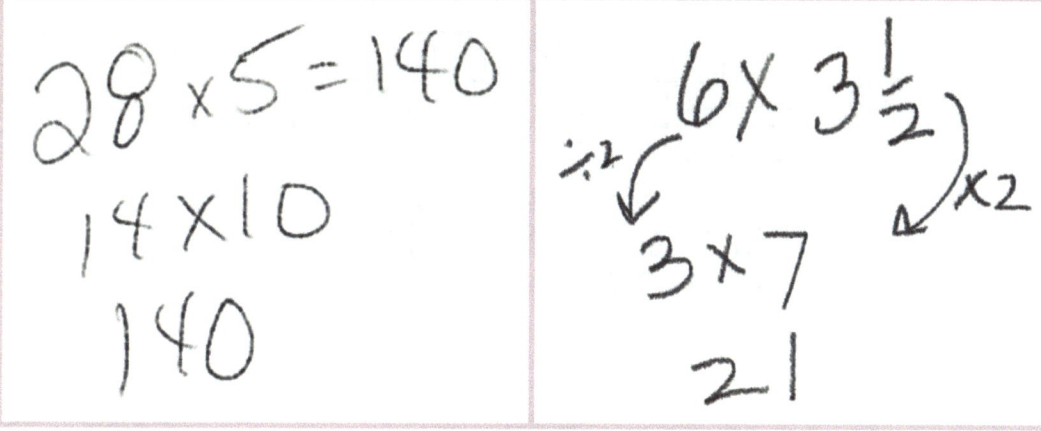

Finding doubles or halves can help students estimate their solutions in order to determine reasonableness. For example, when adding $68 + 63$, a student could estimate with $60 + 60$ or $70 + 70$. In later grades, a student finding the product of 44.68×1.97 can think about 45×2, reasoning that the product should be around 90. Doubling can have other nifty advantages. For example, you could double both the dividend and the divisor in $10 \div 2\frac{1}{2}$ to make the easier $20 \div 4$ or a problem like $135 \div 4\frac{1}{2}$ could be changed to $270 \div 9$.

Also know that learning to double or halve numbers naturally reinforces other strategies and concepts such as decomposing numbers and the distributive property of multiplication. This series is focused on computational fluency, which is part of a larger procedural fluency made up of all sorts of things—comparing fractions, conversions, and even solving systems of equations. The point is that utility with doubles and halves comes from basic fact fluency and builds toward this greater procedural fluency.

The doubling and halving reference page provides an overview, important representations, connections to fluency, and actions to avoid. It can be downloaded and used for reference in planning, teaching, and discussions with colleagues and families.

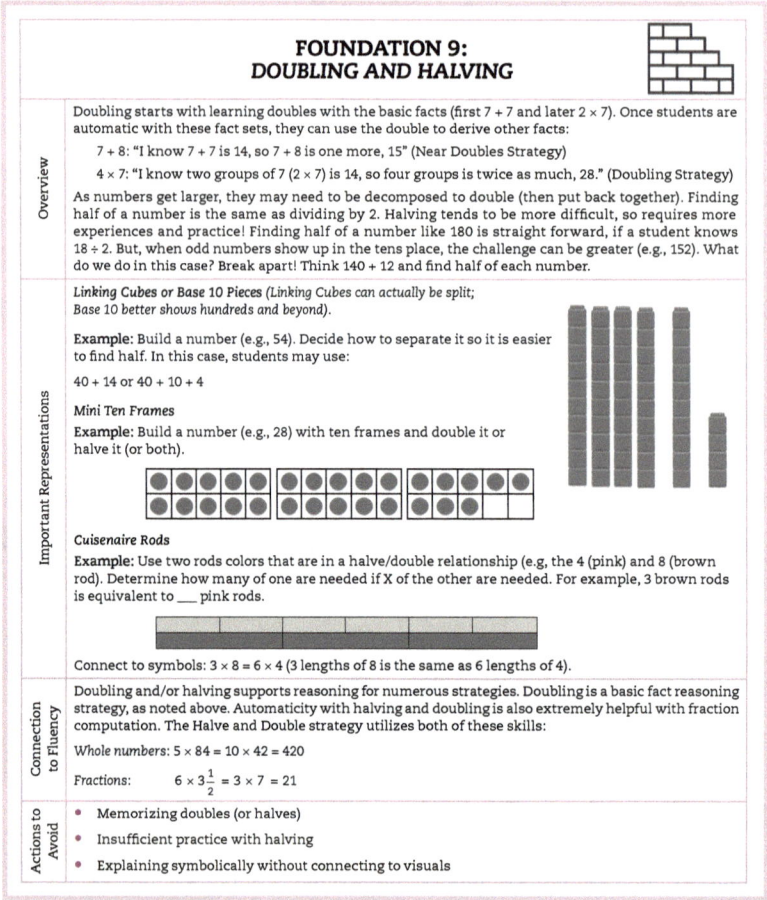

online resources ↳ Available for download at **https://online qrs.ly/psf6a5o**

ASSESSING DOUBLING AND HALVING

When assessing doubling and halving, the question is more nuanced than, "Can this student find half of a number?" The question is, "For what types of numbers is a student able to find half of a number?" This focus helps you to know how to continue to develop this foundation so that they eventually can double and halve more difficult whole numbers as well as fractions and decimals.

DOUBLING AND HALVING LOOK-FORS

First be sure that your students understand doubles and halves. You might pose a dot image and have them create an example that doubles (or halves) the amount, or you can pose a pair of dot images wherein the second image has twice as many dots—perhaps in a different color—to highlight the change. Assuming they're in a good place with the concept, move to basic fact strategies for doubles and halves. Look for students who know (with automaticity) the facts in those sets. Certain doubles, such as 8 + 8 or 9 + 9, can be more difficult than others, such as 3 + 3 or 4 + 4. Look to see that they can transfer basic doubles facts to related multiples of 10 (e.g., 8 + 8 and 80 + 80). Conceptual understanding, basic fact sets, and transference are the bedrock you need to establish before you move into decomposing to double or halve. Look-fors include students being able to

- demonstrate automaticity with addition and subtraction doubles facts (e.g., 8 + 8 = 16 and 16 – 8 = 8),

- demonstrate automaticity with doubles and related division (halves) facts (e.g., 2 × 9 and 18 ÷ 2),

- easily double numbers that don't involve regrouping (e.g., 40, 72, 125),

- easily halve numbers that have all even digits (e.g., 80, 120, 280), and

- decompose a number into "simple" parts in order to double or halve (e.g., break 74 into 60 and 14 to halve).

QUICK ASSESSMENTS FOR DOUBLING AND HALVING

Ask the student what numbers are easy for them to double. Have them give a few examples and explain how they know the double. Extend some of their examples by asking them to double related numbers (e.g., if they double 4, ask them to double 40, 44, or 400).	Show that 97 can be broken apart in different ways, such as 80, 10, and 7; 90 and 7; 80, 15, and 2; 86, 7, and 4; and so on. Note that one or more should be easier parts to double. Then, ask the student which would be a good way to break apart 97 if you were trying to double it.	Find out if a student can decompose and recompose when doubling. Ask them to double 20 and to double 6. Then ask them to double 26. How were they able to double 26? Repeat with a different number.
Give students some easier numbers to halve (10, 18, 28, 46, 62, and so on). Ask them to tell you half of each and how they found half.	Provide a few easier numbers to double: 10, 20, 40, and 50. Ask the student to double each number and show how they found the double.	Ask a student to find half of numbers such as 32, 54, or 76. If they are unable to find half, show them a strategy (e.g., 32 can be decomposed into 20 and 12 to find half of each). Then, see if they can apply this strategy to another number.

EXPLICIT INSTRUCTION FOR DOUBLING AND HALVING

For many adults, strategies for doubling and halving numbers were not taught. You can change this for your students. Have students build models and representations before you explicitly show them how to break numbers apart and recompose them. Intentionally record numbers and their halves or doubles so that patterns appear. Make as many connections as possible during instruction and practice with doubles and halves. Consider where you insert lessons and experiences to support major topics in your grade, for example, simplifying fractions or finding equivalent fractions.

NOTES

_____

_____

_____

_____

_____

ACTIVITY 9.1
CUISENAIRE PROOFS

Cuisenaire rods and a meter stick hold great potential for proving that a number is double that of another.

Students can build numbers with the rods and measure their length in centimeters (cm) with the meter stick. The meter stick allows them to work with larger numbers. This activity is especially useful for numbers that are more challenging to double, such as 38, 49, or 87. To start, have students measure and chart the length in cm of each Cuisenaire rod. Their data will match what's shown in the image.

TEACHING TAKEAWAY

Typical rulers with centimeters on one side or centimeter grid paper can be good substitutes for meter sticks if you don't have them.

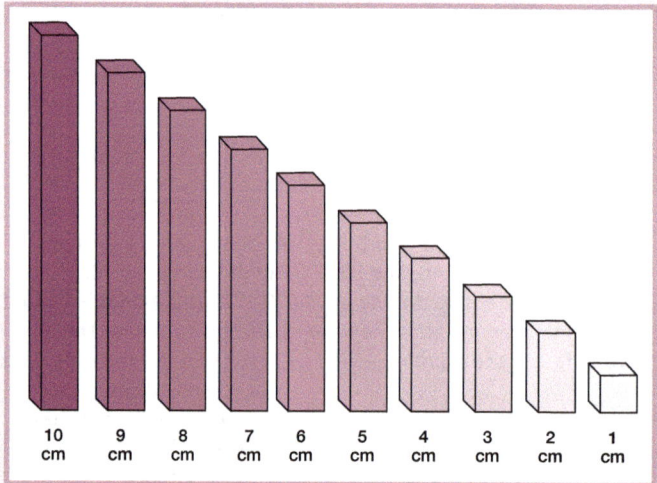

Then, ask them to determine how long two rods of the same color would be. Obviously, they'll find that the length is double. Then, introduce the idea of building multicolor rod pairs such as an orange and a brown or a yellow and a blue. Have them determine the length of these rods. The orange and brown would have a length of 18 cm. The yellow and blue rod would have a length of 14 cm. Then, ask them to predict how long the multicolor rod pair would be if there were two of them (i.e., if it was doubled). After predictions are recorded, have them make models of those rods with two orange and two brown or two yellow and two blue. As you know, the respective lengths will be double (36 cm and 28 cm) respectively.

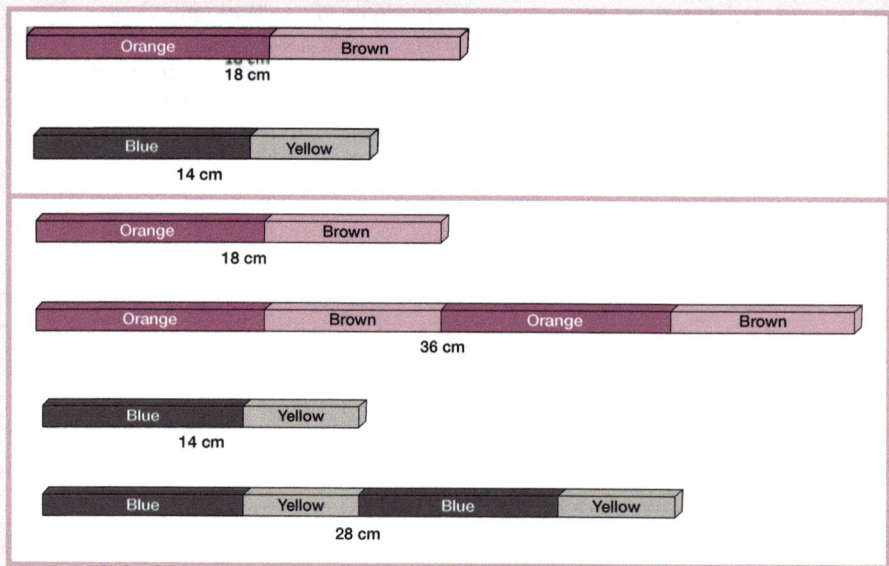

Talk with students about patterns they notice when doubling or the strategies they used to make their predictions (i.e., that they can double both parts separately or double the whole multicolor pair). Have students create their own examples. See if students' predictions become more accurate and what numbers they choose and double. Do they stick with relatively easy numbers to double, such as 20, 30, or 52? Note that this instructional activity can be used in tandem with the next as a way of supporting accuracy and explanation.

ACTIVITY 9.2
DECOMPOSE AND DOUBLE

Doubling 48 by breaking it into 40 and 8, doubling both parts, and recomposing the doubles is a great strategy for doubling more challenging numbers. You want your students to acquire the strategy, but you can't make it a set of steps that they do not understand. First, you want them to see that it can be done. You want them to notice patterns. Then, you want to talk about what are "good" parts and what aren't when breaking apart a number. For example, breaking 48 into 47 and 1 or 37 and 11 probably doesn't help much to find a double.

To begin, use a number such as 48 and ask students to come up with ways to break it apart. Call attention to examples that are easier to double, such as 40 and 8, 42 and 6, or 44 and 4. Have students predict if doubling the parts of a number and putting them back together will be the same as simply doubling the number. Have them break off into small groups to justify their thinking. Bring the class back together to discuss their findings.

After discussion, invite students to prepare a written explanation of how to decompose 48, showing how that number can be doubled by breaking it into parts, doubling those parts, and putting them back together. Repeat the process with other examples.

TEACHING TAKEAWAY

Make calculators available to students to support accuracy. You don't want a powerful idea compromised by faulty computations.

	48	40 + 8	42 + 6	44 + 4
Double it ⟶	96	80 + 16	84 + 12	88 + 8

	54	50 + 4	40 + 14	25 + 25 + 4
Double it ⟶				

	27	20 + 7	25 + 2	21 + 6
Double it ⟶				

	115	100 + 10 + 5	100 + 15	99 + 16
Double it ⟶				

The top of the image shows an example of how you can record the number, its parts, and the corresponding doubles. Recording in this way can help students see the relationships. Below the line are some other numbers you might use with students after the initial experience. You can choose to decompose the numbers in different ways for them (as shown) or let them do it on their own. Do note that in the middle (27) and bottom (115) examples, there are decompositions (19 + 8 and 99 + 16) that don't lend themselves to doubling. Include these examples and engage students to discuss what makes these problems harder to double. Invite students to identify decompositions that are easy to double and ones that are not and to generalize their thinking (e.g., this could be a journal reflection).

ACTIVITY 9.3
BREAKING FOR HALF

When numbers have an odd digit (e.g., in the hundreds or tens place), they are more challenging to halve (e.g., 38, 56, or 176). To find half, as with doubling, one strategy is decomposing the numbers into convenient parts. For example, to find half of 56, one might break it into 40 and 16, finding halves of 20 and 8 before putting them back together. Another way they might think about it is 50 and 6, finding halves of 25 and 3. This idea of decomposing to halve a number doesn't come about organically for many. It has to be taught and practiced (see Activity 9.7).

In this activity, you pose a number that can be challenging to halve, such as 56. Ask students to decompose the number in a variety of different ways independently or with a partner. Bring the group together to share and record a dozen or more examples. Some students might share a decomposition with three parts, as shown on the right in the example. This is fine! Be sure to record a few of those, too, if they come up.

$$56$$

50 and 6   48 and 8

             25, 25, and 6

   55 and 1   40 and 16

             20, 20, and 16

       44 and 12

53 and 3   30, 20, and 6

      10 and 46

           10, 10, and 36

   52 and 4

       44 and 12

39 and 17

    24 and 32   54 and 2

After recording the decompositions that students share, have them break into groups to identify which decompositions are easy to halve. Remind them that both numbers in the decomposition have to be easy to halve. For example, 40 and 16 are probably easy to halve, whereas 38 and 18 might not be. Be sure to have them record the halves of each decomposition. After their group work, bring the class together again to share their findings. Know that each group might not find the same decompositions to be easy.

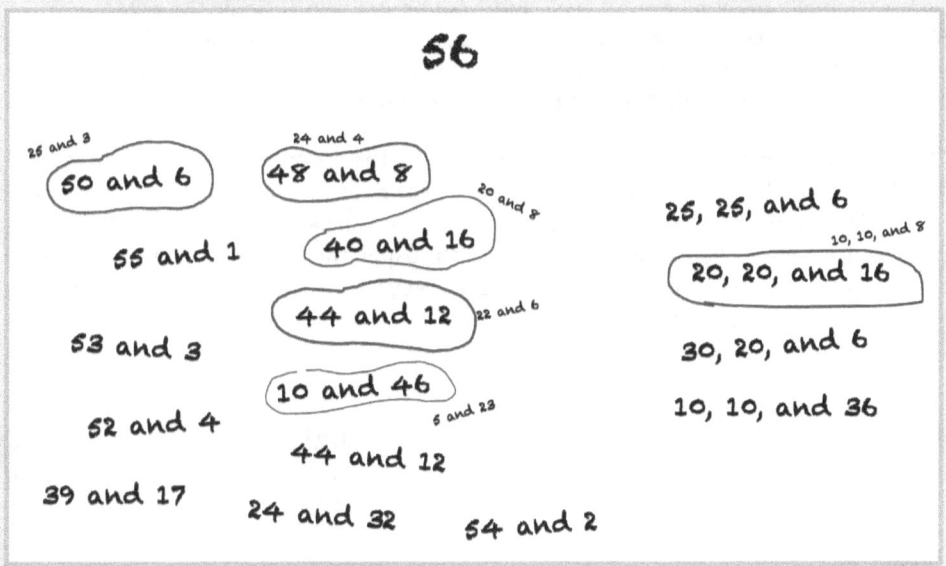

Circle the decompositions they identify, have them share the halves of those parts, and record the halves. In the example, you see that 48 and 8 were easy to halve. It is circled and the halves of each part (24 and 4) are recorded.

Once students have found their favorite decomposed set, ask them to determine that the sum of the halved parts are all equal (28). Then, ask them what they think half of 56 is. Some may be unsure, while others will postulate that if you put the halves of the decomposed parts together, that would also be half of 56. Let students explain their thinking to one another. Avoid asking if it always works right away, instead waiting for them to potentially offer a new example. After a discussion is had, ask students to check their work by finding half of 56 by adding 28 twice, subtracting 28, or dividing by 2. Explore new numbers. You can have a list ready or ask them to identify numbers that are hard for them to halve.

ACTIVITY 9.4
CAN YOU HALVE AN ODD NUMBER?

This activity is about halving odd numbers such as 35, 43, or 77. It should be explored after students show skill with finding half of even numbers and demonstrate understanding of one-half. This lesson leads to a powerful strategy that opens up halving to essentially any number, be it even, odd, fraction, or decimal. Begin the experience by having students find half of different even numbers, some easy to halve (e.g., 60) and some a bit more challenging to halve (e.g., 74). Solicit a number or two from them and talk with the class about how they can halve those numbers. Ask them what they notice about those numbers. Listen to their ideas before sharing (if it doesn't come up) that these are all even numbers.

Ask students, "Can an odd number be halved?" Ask them to justify their arguments using the numbers 35, 43, and 77. Know that some will say that odd numbers cannot be halved based on what they learned about even and odd numbers in first or second grade. Others might already have ideas about how this can be done but will need a tool to help them show it. Linking cubes or ten frames are good options. You can give the experience some "wow" factor by using foam counters and letting students cut *one* counter in half. They'll be amazed that you let them cut the counter. More importantly, they won't forget that halving a number can be done!

You can halve 35 into $17\frac{1}{2}$!

Number	Half
35	$17\frac{1}{2}$
43	$21\frac{1}{2}$
77	$38\frac{1}{2}$

ACTIVITY 9.5
PROMPTS FOR DOUBLING AND HALVING

Use the prompts below as opportunities to develop understanding of and reasoning with finding doubles or halves. Have students use representations and tools to justify their thinking, including base-ten models, number lines, number charts, and so on. After students work with the prompt(s), bring the class together to exchange ideas. Remember that these could be useful for collecting evidence of student understanding.

- Jaxon says he can halve 56 by finding half of 40 and 16 and putting them together. Becca says she finds half by thinking of half of 50 and 6 and putting those together. Bri says she finds half of 10 five times and then half of 6 and puts all of those together. Do each of their ways work? Whose strategy do you like best?

- Scott "just knows" that half of 20 is 10. He doesn't understand why you might have to break apart a number, find half of the parts, and then put those parts back together. What would you say to Scott?

- Asha says that some numbers are easier to double if you break them, double the parts, and then put them back together. Do you agree with her thinking? Give an example of a number that's easier to double by thinking of it this way.

- Some numbers are easy to double (10, 24, 33, or 52). Why are these numbers easier to double? Give three examples of other numbers that are easy to double.

- Mitch says that no matter what number you double, you'll always get an even number. Do you agree or disagree with Mitch? Use examples to show your thinking.

- Sandra says it's easy to find half of 20, 40, 60, 80, and 100. But she has trouble finding half of 30, 50, 70, and 90. What advice would you give her?

- You see that 32 is double 16, 52 is double 26, and 72 is double 36. What are the doubles of 46 and 56? What pattern do you notice in the doubles?

- You see that 9 is half of 18, 19 is half of 38, 29 is half of 58, and 39 is half of 78. What pattern do you notice in the halves? What would you say 49, 59, and 69 are half of?

QUALITY PRACTICE FOR DOUBLING AND HALVING

Quality practice engages students in meaningful and varied opportunities to improve their skills—in this case, their skills with doubling and/or halving. Opportunities for reflection strengthen the impact of practice. On the contrary, some practice works against developing competence and confidence. Practicing a skill or concept before students have a good understanding of it can damage student progress. Grading student practice then deepens uneasiness and apprehension. Timing practice puts undue pressure on students to do things correctly and quickly. Public practice, such as the antiquated "Around the World," holds the potential for public shaming. And while games can be good for engagement, playing competitive games can be off-putting. Know what your students enjoy and work that into practice. Be on the lookout for practice that causes frustration and anxiety. If it causes either, it's not productive.

ACTIVITY 9.6

Name: "High/Low Doubles"

Type: Routine

About the Routine: High/Low is a game for practice with finding doubles. The game is a take on an old card game in which players determine if the next card in the deck will be higher or lower than the card showing. This version is played with a deck of number cards (0–100). A card is revealed, students identify the double of the number, and they predict if the next number/double revealed will be higher or lower than the current one. In this routine, your students play as one team trying to beat the dealer (you).

This version of the game uses numbers 0–100. You can change the range from 0–50, 0–25, or even 100–200.

Materials: Deck of number cards (0–100)

Directions:
1. Tell students that all of the cards in the deck are between 0 and 100.

2. Reveal the first number and ask students to double it. Briefly discuss how they found the double.

3. Ask students to predict if the next number card (and consequently its double) will be more or less (higher or lower) than the current card. Have the class come to an agreement on their prediction.

4. The next card is revealed. Students share what the double of the number is and briefly discuss how they know the double.

5. The game continues until a prediction is incorrect or until the class wins (six correct predictions).

The example shows a game underway. The first card was 22. Students agreed that double 22 was 44, sharing why briefly. They agreed on the prediction that the next card would be higher, and it was (35). They now had one correct prediction. Then, they had to share its double (70) and how they knew it was the double. They predicted the next card would be higher. So far, they have two predictions correct. The image on the right shows how the teacher recorded the cards and their doubles on the board while the game was played.

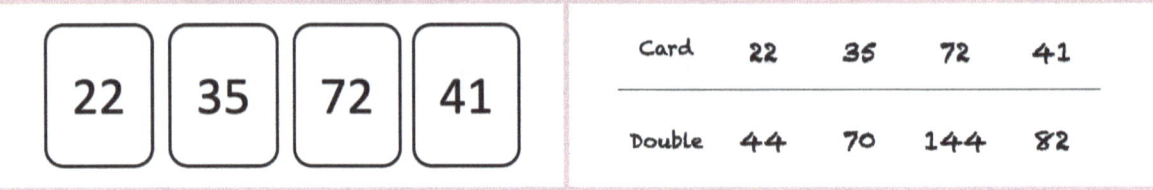

Card	22	35	72	41
Double	44	70	144	82

online resources ⇲ A set of cards from 1 to 100 can be downloaded at **https://qrs.ly/psf6a5o**

ACTIVITY 9.7

Name: "String of Halves"   **Type:** Routine

About the Routine: String of Halves is an opportunity to practice strategies for finding half, as described in Activity 9.3. To be clear, the idea of breaking apart a number, finding half of those parts, and then recomposing the partial halves must be understood before this routine should be used. The routine works with any number. Simply choose a number that's more challenging to half (36, 58, 76, and so on). These numbers usually have an odd number of tens or hundreds.

Materials: Prepare a set of numbers that are intentionally related.

Directions: 1. Post a set of three or more related numbers as shown. In the example, the teacher selected 36 and broke it into 20 and 16.

2. Ask students to mentally find the halves of as many of the numbers as possible.

3. If students are unable to find all of the halves, have them talk with partners about how they can use the relationships between numbers and their known halves to find the unknown half. If students find all of the halves, have them discuss with partners their strategies for finding the halves and how the numbers (and halves) are related.

4. Optional: To extend this situation (when all halves are known), have students generate a new number and its half that is related to the set.

In the example below, the teacher posed the string on the left (20, 16, and 36). Students thought, discussed, and shared the halves of 20 and 16, which the teacher recorded as shown. Then, they were asked to figure out half of 36. They reasoned it to be 18: "You can put the halves together because 20 and 16 are put together to make 36." The right side shows how the teacher extended the routine by asking students to think of another way to decompose 36, checking to see if their halving strategy would still work. Students came up with 24 and 12.

20	16	36		?	?	36

20	16	36		24	12	36
10	8			12	6	

ACTIVITY 9.8

Name: *For Keeps (Doubles or Halves)*   **Type:** *Game*

About the Game: This game is based on the *For Keeps* game that appears in Activity 7.10 and in the *Figuring Out Fluency* anchor book (Bay-Williams & SanGiovanni, 2021, p. 120). It appears in this module, illustrating that many games in other modules or books can be modified to practice doubles or halves by changing the directions.

In the directions, #4 is a reminder that you want students to play all eight rounds regardless of how quickly they find the numbers or points they're going to keep so they get more practice.

Materials: *For Keeps* game board; digit cards (0–9), playing cards (queens = 0, aces = 1; remove tens, kings, and jacks), or a 10-sided die

Directions:
1. Players take turns generating two numbers with a die or cards.
2. Players double their number.
3. The player decides whether they want to keep the double as one of their four final scores. Players can't change their mind on later turns.
4. Players must take eight turns, regardless of how many turns it takes for them to find the four scores they want to keep.
5. After eight turns, players find the sum of their kept doubles. The sum is their score.
6. The player with the highest score wins.

> **TEACHING TAKEAWAY**
> Modifying familiar and favorite games can help students focus on mathematical skills rather than learning the rules and strategies for a new game.

The game board on the left is for doubling. The game board on the right shows the modified version for finding half, as mentioned above. As with other halving activities, you can restrict students to only halving even numbers until finding half of an odd number is appropriate.

For Keeps

Directions: Take turns making a number. Double the number and decide if you want to keep the double as your score. You can only keep four of your doubles. After eight turns, add the doubles you kept to find your score. High score wins.

Round	Number	Double KEEP	Double DON'T KEEP
	43	86	
1			
2			
3			
4			
5			
6			
7			
8			
SCORE --------> Sum of the four products I kept			

For Keeps

Directions: Take turns making a number. Halve the number and decide if you want to keep the half as your score. You can only keep four of your halves. After eight turns, add the halves you kept to find your score. High score wins.

Round	Number	Half KEEP	Half DON'T KEEP
	62		31
1			
2			
3			
4			
5			
6			
7			
8			
SCORE --------> Sum of the four products I kept			

online resources This resource can be downloaded at **https://qrs.ly/psf6a5o**

ACTIVITY 9.9

Name: The Splits   **Type:** Game

About the Game: *The Splits* is a game of reasoning and luck that practices halving or splitting numbers. The object of the game is to be the first player to fill your boxes with numbers and their halves in order of least to greatest. It seems simple. But players will have to think carefully about where they place their numbers! As students first work with halves, restrict the numbers they can halve to even numbers. You can choose to allow players to rearrange the digits or not; for example, a 2 and a 7 can be used for 27 or 72.

> **TEACHING TAKEAWAY**
> Reverse the game to practice doubling. To do this, have them make a number and record it on the top, then double it and record the double below.

Materials: Digit cards (0–9) or playing cards (queens = 0, aces = 1; remove tens, kings, and jacks), *The Splits* game board, *The Splits* recording page

Directions:
1. Players take turns generating a two-digit number. Note that three- and four-digit numbers are options, if appropriate.

2. Players record that number in one of the six boxes in the bottom row.

3. Players halve the number and record the half in the box above the number they generated.

4. Each subsequent number and its half must be placed in a box so that all numbers are in order from least to greatest.

5. If an odd number is generated or a half cannot be placed, the player loses their turn.

6. The first player to fill their boxes wins.

In the example below, the player first made 48. They placed it, halved it, and recorded it as shown. On their next turn, they made 34 and found half. On their third turn, they had a 4 and 2. They have no place to put 42 and its half, because it would be placed to the left of 48, where there is no space. So, they lost their turn.

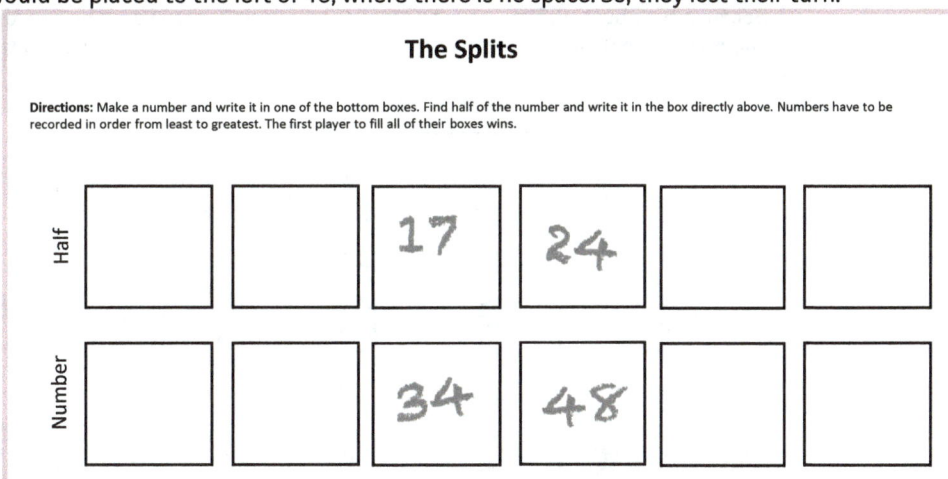

 A blank game board can be downloaded at **https://qrs.ly/psf6a5o**

ACTIVITY 9.10

Name: First to Seven　　　　　　　　　　**Type:** Game

About the Game: In this game, you want to be the first to make seven even numbers (i.e., numbers with whole-number halves). It is a simple game with a bit of chance that makes it fun for students of all ages. You can adjust the rules of the game so that students are allowed to rearrange the digits or you can change the game from two-digit numbers to three-digit numbers.

Materials: One deck of playing cards (aces = 1, queens = 0; tens, jacks, and kings removed) or one set of *First to Seven* cards per player

Directions:

1. Player 1 begins by flipping two cards to make an even two-digit number. If Player 1 draws two odd numbers, the play passes to Player 2. If Player 1 can create an even number, they state what half of the number is. Player 2 confirms that Player 1 is correct. If Player 1 is incorrect, it's the next player's turn.

2. If correct, Player 1 can continue or pass their turn to the next player.

3. If Player 1 goes again, they have to half another even number. They tell what the half is. If correct, Player 1 chooses again if they want to pass their turn or try for a third half in a row.

4. Any time a player chooses to play again, they must make an even number and correctly halve it. If they don't, they lose all of their progress, shuffle their cards, and start over. Once a player chooses to pass, they get to stay at that mark and will come back to that spot if they bust on another turn.

Below, you see Player 1's turn. On their first turn, they made 62 (half is 31) and chose to go again. Then, they made 30 and chose to go again. Then, they made 42. They were up to three halves (31, 15, and 21). Then they made 71, which can't be halved, so they busted. Because they busted, they have to start over and go back to 0 halves.	Below, you see Player 2's turn. On their first turn, they made 18 (half is 9) and chose to go again. Then, they made 50 (half is 25). With two doubles (9 and 25), they decided to pass. When play came back to them, they made 39. Even though 39 doesn't have a whole-number half, they kept the two doubles they already had because 39 was on their first play of a turn.

Player 1's cards:

- Turn 1 – 1 double (goes again): **6 2**
- Turn 1 – 2 doubles (goes again): **3 0**
- Turn 1 – 3 doubles (goes again): **4 2**
- Turn 1 – busted (has to start over): **7 1**

Player 2's cards:

- Turn 1 – 1 double (goes again): **1 8**
- Turn 1 – 2 doubles (passes): **5 0**
- Turn 2 – busted (Still 2 doubles): **3 9**

A deck of cards for this resource can be downloaded at **https://qrs.ly/psf6a5o**

ACTIVITY 9.11

Name: 50-150   **Type:** Game

About the Game: In this game, players practice doubling and get points when their double is between 50 and 150. But there is a catch: If they don't double their number correctly, they can't earn the points and if the correct double is less than 50 or more than 150, they lose a point! The first player to earn 10 points wins the game.

For games with more than two players, adjust the points so that a game doesn't take as long. Alternatively, students can play until the deck is gone and the player with the most points at the end wins. Another alternative for the game includes using kings and/or jacks as wild cards (the player chooses any digit they want the card to represent). You could choose to make numbers between 50 and 150 worth two or three points.

> **TEACHING TAKEAWAY**
> Keep games fresh by changing the rules and conditions, creating different challenges and renewing interest.

Materials: Four decks of digit cards (0–9) or playing cards (queens = 0, aces = 1; remove tens, kings, and jacks)

Directions:
1. The first player flips over two cards to make a two-digit number.

2. The player doubles the number and their opponent checks for accuracy using a calculator (e.g., 35 + 35 or 2 × 35).

3. If the player is correct and their double is between 50 and 150 (including 50 and 150), they get a point. If they are correct but their double is less than 50 or more than 150, they don't get a point.

4. If the player is incorrect, they lose a point.

5. The goal of the game is to score 10 points.

The example shows a game with two players. Player 1 drew a 7 and a 2 and chose 27. They said the double was 54; their opponent checked and confirmed their answer. Since 54 is between 50 and 150, Player 1 got one point. Player 2 drew a 6 and a 5, chose 65, and correctly found the double of 65 and got one point. On their next turn, Player 1 said the double of 91 was 192, which was incorrect, so they lost one point. Player 2 then found the double of 46 correctly and got a point. After two turns, Player 1 has no points and Player 2 has two points.

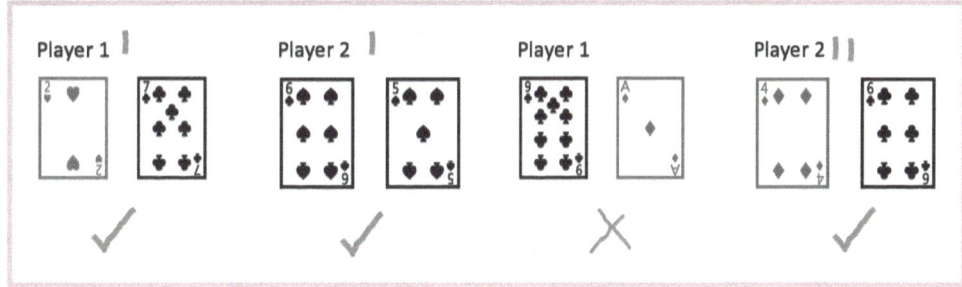

ACTIVITY 9.12

Name: Halve it Up
Type: Game

About the Game: *Halve It Up* is a simple one-player game that functions as a center for students to practice halving numbers. This center does not halve odd numbers. In fact, the whole premise of the experience is to avoid generating an odd number because that ends the game. Students generate a number and find half. They record their ideas and use a calculator (e.g., $x \div 2$) to confirm they are correct.

Materials: Digit cards (0–9) or playing cards (queens = 0, aces = 1; remove tens, kings, and jacks), *Halve It Up* recording sheet (optional)

Directions:
1. A student flips over two cards to make a number. Cards can be rearranged (e.g., 4 and 8 can be 48 or 84).

2. The student determines half of the number and records it. The student uses a calculator to check if their half is correct.

3. The student then flips over two new cards, finds half, records, and checks.

4. The student continues making numbers and finding half until a number can't be halved into a whole number or an incorrect half is found.

5. The game is over when the player has gone through the entire deck, creates an odd number (can't be halved), or is incorrect.

6. The goal of the game is to make it completely through the deck.

> **TEACHING TAKEAWAY**
> This activity comes with a recording sheet, available on the companion site. However, students can easily record their work in their mathematics journals or on a piece of paper.

The image shows a student playing Halve It Up. On their first turn, they made 68. They figured half to be 34 and checked to confirm. Their check mark shows they are correct. Then, they had a 6 and a 3 and decided to make it 36 and found half to be 18. Then, they found 42 and half that was 21. On their fourth turn, 31 is odd (so is 13) and the game is over. They didn't make the goal of getting through the entire deck.

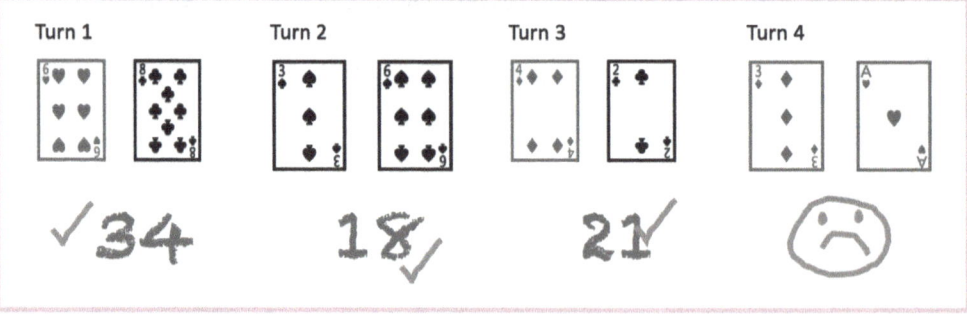

Halve It Up

Directions: Make a two-digit number. Find half. Check to see if you are correct. Keep going until all of the cards are gone, you can't find half of a number, or your guess is incorrect.

My Number	Half	Correct?	My Number	Half	Correct?
		Yes No			Yes No
		Yes No			Yes No
		Yes No			Yes No
		Yes No			Yes No
		Yes No			Yes No
		Yes No			Yes No
		Yes No			Yes No
		Yes No			Yes No
		Yes No			Yes No
		Yes No			Yes No
		Yes No			Yes No
		Yes No			Yes No
		Yes No			Yes No
		Yes No			Yes No

 This resource can be downloaded at **https://qrs.ly/psf6a5o**

Computational Estimation

COMPUTATIONAL ESTIMATION OVERVIEW

The difference between estimation and computational estimation is that *estimation* has to do with estimating a single quantity ("How many jellybeans do you estimate are in this jar?") whereas *computational estimation* comes into play when we estimate the results of operations ("The product will be about ___"). *Computational estimation* refers to estimates used with computation. That may sound obvious, but it can be confused with estimating how many candies are in a jar, which is *not* computational estimation. There are three strategies for computational estimation (Bay-Williams & SanGiovanni, 2021):

- **Rounding:** Using nearby benchmark numbers that make the computation easy; for example, 219 – 94 could be rounded to 220 – 90 or 220 – 100. Estimates are 130 or 120.

- **Front End:** No rounding is required, simply look at the largest digit; for example, 219 – 94 would be 200 – 90. The estimate is 110.

- **Compatible Numbers:** This strategy adapts the numbers so they work well together. For 219 – 94, this could be going up a little with both to 225 – 100 (an estimate of 125). This strategy is the best for division (rounding rarely helps). For example, consider 685 ÷ 8. Rounding 685 to 690 or 700 is not helpful, but changing 685 to 640 is helpful. The estimate is 80.

- **Range:** Range uses rounding down and rounding up to set the range in which an estimate can fall. This can be a very useful strategy for students who have trouble choosing compatibles. For 219 + 425, the lower bound is 600 (front-end estimation), the upper bound could be 220 + 430 (650). For 143 × 27, the lower bound could be 100 × 20 and the upper bound could be 200 × 30. The upper bond is much greater. Finding the range is flexible and the key is not to get bogged down in setting the range but to find easy numbers to compute.

Computational estimation is hard to teach because students can be uncomfortable with the numbers, strategic options, and inexact answers. The key is to communicate that the goal is not to take too much time trying to get as close to the actual answer as possible but rather to get a sense of about what the answer will be.

Computational estimation isn't always featured in your mathematics program or instructional resources, and even when it is, it is unlikely to be sufficient—students need to be estimating all the time so that they can decide if their answers are reasonable. This module connects to and builds on Module 1, with a strong focus on thinking flexibly about numbers and how they're related to others. It provides ideas for teaching *how* to estimate and *when* to estimate. Computational estimation can't be practiced enough. Incorporate daily opportunities even if it's in some small way, such as having students estimate a result before doing a problem and reflecting on the reasonableness of that answer. Know that lots of practice and discussion about estimating sums and differences or products and quotients will go a long way in helping your students realize their fluency.

HOW DOES COMPUTATIONAL ESTIMATION CONTRIBUTE TO FLUENCY STRATEGIES?

Is this reasonable? Does my answer make sense? These are the questions you ask yourself when you do mathematics and are questions you want your students to ask themselves! To check for accuracy, students need to have an idea of what their result will be. They must think, "What is reasonable?" A student who adds 38 + 49 and finds a sum of 177 should pause, realizing that 177 is unreasonable. It's too much! They look at their problem and think about 40 + 50 before digging back into the problem to see where they might have gone wrong. Similar experiences play out with subtraction, multiplication, and division. Thus, establishing reasonableness of an answer is reliant on computational estimation.

Computational estimation parallels computational fluency. As noted in the overview, there are reasoning strategies for estimating just as there are reasoning strategies for computing. And like computational strategies, estimation strategies change relative to the numbers within the problem. A student may begin with rounding and switch to front-end estimation. Hence, as with computational fluency, flexibility is critical in developing fluency with computational estimation.

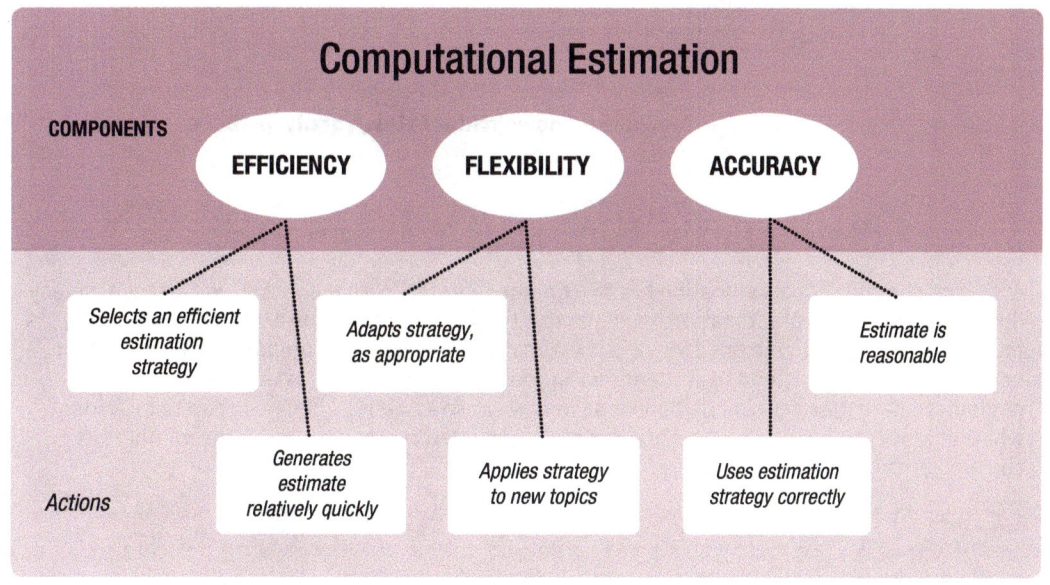

Source: Adapted with permission from D. Dpangler & J. Wanko (Eds.), *Enhancing Classroom Practice with Research behind Principles to Actions,* copyright 2017, by the National Council of Teachers of Mathematics. All rights reserved.

The computational estimation reference page provides an overview, important representations, connections to fluency, and actions to avoid. It can be downloaded and used for reference in planning, teaching, and discussions with colleagues and families.

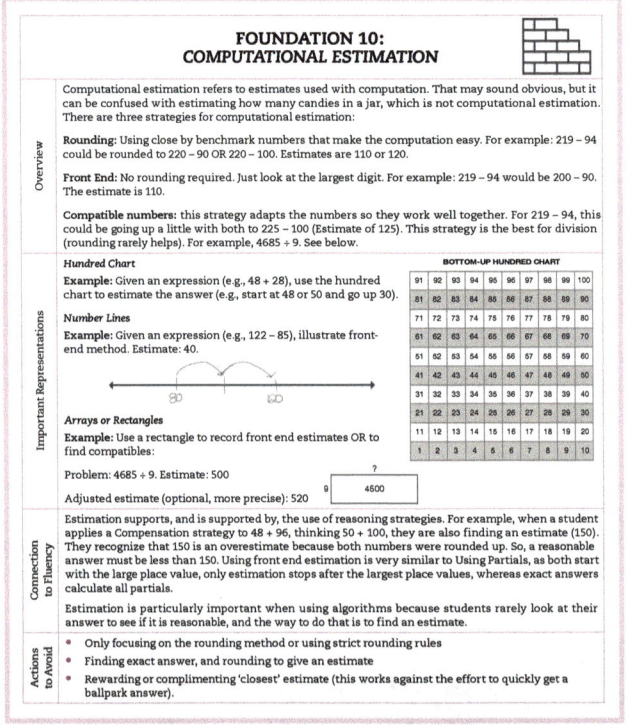

online resources
Available for download at **https://qrs.ly/psf6a5o**

TEACHING TAKEAWAY

Avoid focusing on the closest estimate and instead recognize answers that fall in a reasonable range.

ASSESSING COMPUTATIONAL ESTIMATION

Assessing computational estimation has two unique features. With many skills, we assess accuracy. But with computational estimation, students can try too hard to get the answer or spend too much time on a strategy to get a very close estimate. First, assessing computational estimation needs to look for and attend to students' willingness and ability to get in the ballpark. Second, computational estimation can be done in several ways, so assessing activities need to look for whether students know how to *use* and effectively *choose* a method based on the numbers in the problem.

COMPUTATIONAL ESTIMATION LOOK-FORS

At the core of estimating is understanding number relationships, benchmark numbers, and compatible numbers, so this is a good place to begin assessing student understanding. Look back to Module 1 for ideas about what else you might look for. When enacting computational estimation strategies, listen for students to describe estimates using conceptual language (e.g., "The number is close to 50") versus rule-based language (e.g., "It was a 5, so I rounded up"). As you focus on computational estimation, the use of compatible numbers becomes increasingly important. For example, when adding 177 + 349, a student might round the addends to 180 and 350, but these are not that easy to add. Thinking of compatibles, a student might instead use 180 + 320. To subtract 432 − 165, using compatibles is much easier than a rounding method: A student can change the first number to 465, creating 465 − 165; adjust the second number, creating 432 − 132; or ballpark them both at 450 – 150. Estimating with division requires finding compatibles, which involves recognizing basic fact combinations. To divide 714 ÷ 8, for example, a student can think of the problem as 720 ÷ 8 or 700 ÷ 10. As computational estimation develops, you want to know that your students can

- think flexibly about numbers and their relationships;
- identify compatible numbers (e.g., combinations of 10, 100, or 1,000);
- compose numbers such as tens and hundreds, which guide their number choices;
- demonstrate that they can use each of the four computational estimation strategies;
- demonstrate flexible use of the four computational estimation strategies and justify their approach;
- use estimation to determine reasonableness; and
- demonstrate the fluency actions in the computational estimation visual above.

QUICK ASSESSMENTS FOR COMPUTATIONAL ESTIMATION

The prompts below are a beginning list of ideas for assessing students' readiness to estimate, and their use and flexibility with the computational reasoning estimation strategies. Any prompt can be adapted to other operations. More prompts can be found in Activity 10.5.

Pose a number, such as 84, and ask the student if it's close to 0, 50, 80, 90, or 100. Ask the student how they know what numbers it is close to.	Pose an addition problem, such as 38 + 44 or 177 + 349 (depending on the student's grade level). Ask the student to estimate the sum and explain how they estimated.	Pose a few problems, such as 84 + 78, 36 + 50, and 29 + 45 or 12 × 15, 22 × 9, 45 × 6. Ask the student to identify which problem(s) are good to estimate and which problem(s) they don't need to estimate.
Ask the student to give you examples of good numbers to use when they are estimating with multiplication (e.g., 24 × 9 or 89 × 35). Ask them to tell you why those numbers are good for estimation.	Pose a problem, such as 534 + 278, 71 − 36, 39 × 31, or 522 ÷ 8. Ask the student to estimate the difference in two different ways and explain which way they like the best and why.	Pose a problem and ask the student if the result will be more than or less than a given number. For example, pose 64 + 77 and ask if the sum will be more than or less than 100 or 200.

EXPLICIT INSTRUCTION FOR COMPUTATIONAL ESTIMATION

To teach computational estimation well, you must be familiar with the strategies and representations of them.

Avoid a single "best" approach to estimating (namely rounding). Along that line, it is extremely important that as you teach students that there are always a range of reasonable estimations (there is never a single correct estimate). Teach each method for computational estimation, encourage students to use the method, and then provide many opportunities for them to choose among the methods. Encourage flexibility. And remember, the best way to grow computational estimation is to practice and talk about it daily.

ACTIVITY 10.1
WHAT IS REASONABLE?

Estimating helps determine if an answer is reasonable. But what *is* reasonable? For adults, reasonable is usually clear. We have practiced *reasonable* countless times. For some students, *reasonable* can be a murky term, especially if they perceive mathematics to be a rule-based, procedural subject where the correct answer is the driving purpose.

This activity is designed so that students can explore reasonableness. Students might have a little fun, too! Start out with some fun scenarios to get your students thinking about what might be reasonable for a given situation. Pose a situation and ask them to determine, "Which would be more reasonable?"

WHICH WOULD BE MORE REASONABLE?		
To walk from Chicago to Washington, DC or from your house to the school?	For one person to eat 32 s'mores or 3 s'mores while sitting around a campfire?	For a person to have 25 dogs or 25 fish?
To use fingernail polish or soda as paint for a canvas?	To shoot hoops with a beach ball or bowling ball?	For a person to take a nap in a chair or on a bicycle?

Once students have a better sense of *reasonable*, move to more complex problems. Present problems to students and have them work collaboratively to throw out or narrow down the possible solutions. Students will eliminate the answer that is furthest from the solution and work toward landing on the answer that is most reasonable. For example, a problem may read, "Popsicle sticks come in packs of 115. If I get five packs from Amazon for a school project, approximately how many total popsicle sticks would I have?" Students consider the answer choices of 15, 500, 1,150, and 5,000. They share their reasoning of why an answer option other than 500 would be eliminated, starting with the most obvious. Below are samples of student reasoning:

SAMPLE STUDENT JUSTIFICATIONS	
15	"This is a silly answer. There are more than 15 in one box!"
5,000	"This is *way* too many! If I think of each box having about 100 in each, there is no way there would be 1,000 sticks!"
1,150	"I think this answer is just the numbers put together. It doesn't make sense as 100 + 100 + 100 + 100 + 100 doesn't make it to 1,000."
500	"This answer is most reasonable. If you think about 100 in each box, that is about 500 sticks. And that sounds about right for a school project."

Examples of possible problems to use with students for this activity are in the table below. Hopefully these provide a springboard for other scenarios that could be used for this instructional activity.

EXAMPLE PROBLEMS	POSSIBLE ANSWER CHOICES	MOST REASONABLE ANSWER
The baseball team had 20 baseballs in the bucket. They lose a few as players hit them over the fence. How many balls could be in the bucket at the end of practice?	1, 3, 18, 90	18
Each of the 57 students in second grade are able to bring two guests to the poetry reading. How many chairs should the custodian set up for the event's guests?	59, 110, 572, 5,072	110

EXAMPLE PROBLEMS	POSSIBLE ANSWER CHOICES	MOST REASONABLE ANSWER
It costs $1.45 for an ice cream treat. If the coach wants to buy one for each of the 30 players on his soccer team, roughly how much money will the coach spend?	$4.00, $14.00, $40.00, $400.00	$40.00
LaKau reads every night before he goes to bed. He's read a total of 783 pages in the last five days. About how many pages does LaKau read each night?	150, 310, 400, 900	150

After students have the opportunity to rule out outlandish or unreasonable solutions, it is time for them to practice finding reasonable answers on their own. Provide students with problems similar to those used before, however, the students will now offer a reasonable answer and the justification they used to land on that estimation. Here is the problem: "In the typing game, the student types 32 words in one minute. Around how many words would this student type in seven minutes?" A student might say about 200 words is reasonable, as they are thinking that 32 is close to 30, and 30 multiplied by 7 is 210. They might also have some fun by creating their own sets of answers, including answers that are way off track!

ACTIVITY 10.2
DO I NEED TO ESTIMATE?

Should I estimate my solution? This is a good question throughout the process of computing. It's considered when numbers are first encountered, as a problem is worked, and when solutions are found. Estimation is also considered situationally in problems as the contexts and questions cause one to think about the need for an exact or an approximate answer. This activity is an opportunity to help students learn to pause and ask if they need to estimate a solution.

In this activity, students consider a problem and determine if they can find the exact answer with little effort or if the exact answer will be hard to figure out and they should estimate first. To do this, prepare a collection of expressions suitable to the computations your students have been working with. Write or print them on small strips of paper. Examples of problems from different levels include those in the table below:

Addition Examples		Subtraction Examples	
37 + 10 **	38 + 19	79 – 2 **	91 – 49
9 + 59	97 + 18	61 – 9	32 – 16
77 + 8	50 + 40 **	50 – 10 **	73 – 20 **
1 + 16 **	41 + 38	43 – 8	58 – 39
7 + 56	77 + 87	82 – 7	112 – 76
Multiplication Examples		Division Examples	
17 × 3	13 × 10 **	210 ÷ 7	149 ÷ 15
20 × 4 **	21 × 48	354 ÷ 6	34 ÷ 17 **
39 × 1 **	15 × 32	120 ÷ 6 **	160 ÷ 80 **
73 × 6	40 × 40 **	88 ÷ 4 **	485 ÷ 15
14 × 8	12 × 12	92 ÷ 5	96 ÷ 16

Have students work in partners or triads to examine each problem and determine whether they need to estimate a result before calculating. Then, they sort their problems into those that need estimating and those that don't. As students work, circulate around the room to take stock of their thinking. When most are finished, bring the groups back together to share their findings. Be sure to focus on why certain problems don't need an estimate (because the exact can be found quickly and easily). In the chart, problems marked with asterisks (**) are good examples of problems that students should determine as problems that don't need to be estimated. For example, in a second or third grade class, you would want students to say that they don't need to estimate 50 + 40 because they know the answer or that they can count on by tens quickly.

TEACHING TAKEAWAY

This activity is perfect for revisiting again and again through a routine. To do this, pose two or three problems and have students turn and talk about whether they need to estimate. Then, discuss their conclusions as a class.

ACTIVITY 10.3
JUST ONE

This activity focuses on flexibility in choosing compatible numbers for estimating. For many students, their first forays into estimating sums and differences call for changing both numbers (i.e., addends) into two benchmark numbers. There is value in changing both numbers in a problem into something compatible. Yet, it's often more advantageous—and less complicated—to manipulate only one of the numbers. This becomes more and more true as students become proficient with counting on and back by tens, hundreds, and groups of tens or hundreds.

TEACHING TAKEAWAY

If you anticipate that no students will come up with a certain approach or strategy, you can create worked examples for students to analyze and react to.

This lesson takes aim at introducing students to the notion of changing only one of the numbers to estimate a sum or difference. Present them with a problem and ask them to estimate the sum however they like. Discuss and compare student approaches. Ideally, one student or student group will estimate by changing one number. When that happens, have students compare the approaches, estimates, and efficiency of changing one or both numbers. Be prepared for students to think that the rule or correct way to estimate is to change both numbers.

37 + 47

Change Both Addends	Change One Addend	Change One Addend
40 + 50 = 90	37 + 50 = 87	40 + 47 = 87

Exact
37 + 47 = 84

Don't be surprised if none of your students estimate a sum or difference by manipulating one number. That's why this lesson is valuable! If no students do this, present an example and ask them to compare the approaches.

Introduce the exact sum or difference. Have them grapple with which estimate is closest and why they think that is so. Some students will be quick to recognize the value in estimating by changing one number. Others will show that they are unsure about the value or correctness of the approach. That's OK, too!

To continue the activity, create a collection of problems for partners to work with and rotate through. With each problem, charge students with estimating the sum by changing each addend. After students work through the problems, bring the class together to compare results.

ACTIVITY 10.4
ESTIMATING WITH WORKED EXAMPLES

As noted in the overview, rounding, front-end estimation, compatible numbers, and finding the range are three predominant strategies for computational estimation results. Front-End and Range strategies are excellent options for students who have trouble choosing compatibles because they are more straightforward. Range uses rounding down and rounding up to set the area in which an estimate can fall. For addition, you round both addends down for the lower bound and round both addends up for finding the upper bound. The same approach is used for multiplication.

This activity begins with a worked example so that you can engage students in observing and making sense of the approach. A good prompt might look similar to either of these, depending on the content you teach.

Front-End Worked Examples:

To add 625 + 289, Nico decides to estimate first: 600 + 200 = 800 What did Nico do? Nico wonders, should he stay with his estimate or go up 100. What do you recommend, to stay or go up?	Nico is estimating 92 × 12. He writes this: 90 × 10 = 900 What did Nico do? Nico wonders, should he stay with his estimate or go up 100? What do you recommend, to stay or go up?

Range Worked Examples:

Asia is trying to solve 237 + 339. She thinks it will be hard, but it would help to estimate first. Look at her work. What did she do? Do you think this always works?	Asia is trying to solve 57 × 14. She thinks it will be hard, but it would help to estimate first. Look at her work. What did she do? Do you think this always works?
$237 + 339 =$ more Than $230 + 330 = 560$ Less Than $240 + 340 = 580$	57×14 more than $50 \times 10 = 500$ less than $60 \times 20 = 1,200$

Ask students, "Is this a good way to estimate?" Be prepared for students to say: "No, this isn't how you estimate." After discussion, reveal that it *is* a way to estimate and they can choose it when it is more efficient than other methods they have learned. After discussing the worked example, challenge students to estimate new problems (e.g., using the Front-End strategy or finding the range by writing more-than and less-than equations).

You can build a routine from this instructional activity once your students show that they are comfortable with the strategy. To do this, pose some problems for partners to solve. For the Front-End strategy, one partner finds a front-end solution and the other partner decides to stay or go (then switch). For range, one partner finds the upper limit and the other partner finds the lower limit. Have partners share, then discuss as a class what estimates/ranges were found.

Finding the range changes when working with subtraction and division. For subtraction, the lowest number you might get comes from rounding the minuend (first number) down (start low) and rounding the subtrahend (second number) up (the most you can take away). Then, to find the upper bound, you round the minuend up (start high) and round the subtrahend down (the least you can take away). For division, students find the two compatibles' dividends. Take, for example, 468 ÷ 7. Use both 420 or 490, showing the range as between 60 and 70.

ACTIVITY 10.5
PROMPTS FOR COMPUTATIONAL ESTIMATION

Use the prompts below as opportunities to develop understanding of and reasoning with the strategy. Have students use representations and tools to justify their thinking, including base-ten models, number lines, number charts, and so on. After students work with the prompt(s), bring the class together to exchange ideas. Remember that these could be useful for collecting evidence of student understanding.

- How would you describe finding a reasonable answer to a friend?

- Gina estimates a problem to have a sum of 120. What do you think the problem might have been? Explain your thinking.

- Would you expect the difference of 701 – 399 to be more than or less than 200? Explain your answer.

- Raj says that 24 rounds to 20, but sometimes it's better to think about it as 25. Give an example of when it would be good to round 24 to 20 and an example of when it would be good to think about it as 25.

- Explain how rounding and estimating are similar and different. Give an example of your thinking.

- Amy estimated the sum of 143 + 489 and 157 + 508. She was surprised that she had the same estimate for both sums. What do you think that estimated sum was? How did she think about each problem to get to that estimated sum?

- Kai estimated the difference of two three-digit numbers to be 150. What do you think the original problem might have been? How might Kai have estimated that problem?

- Create three problems that you don't need to estimate the sum or difference for. Why don't you need to estimate those sums or differences?

QUALITY PRACTICE FOR COMPUTATIONAL ESTIMATION

To get the most out of practice, introduce routines, games, and centers through a lesson. Play a game in which the class is a team and you are the opponent. Use a center as an instructional task for a small group. As you introduce activities, highlight the questions you want students to ask themselves. Use think-alouds to help develop students' metacognition and processing of the activity and—more importantly—of the mathematics. Teaching students how to work with an activity and how to think about mathematics will pay big dividends!

ACTIVITY 10.6

Name: "Is It Reasonable?"   **Type:** Routine

About the Routine: This routine is different because problems are already evaluated. Instead of estimating a result, students are given a statement and asked to consider whether it is reasonable. They will call on their emerging estimation skills to reason through the prompts. It works with any operation or set of numbers. Two or three statements are plenty for a practice experience. A good source for problems can come from daily instruction. To do this, take note of problems that students have difficulty with. You can then use those problems in this routine in the following days.

Materials: Two or three "Is about ___" statements (see the chart below)

Directions: 1. Pose a prompt to students and have them think independently about whether it is reasonable and why.

2. Have partners share their ideas.

3. Bring the whole class together to ask how many students think the statement is reasonable and how many think it isn't reasonable.

4. Have students discuss their thinking.

5. Clearly establish if the statement is or isn't reasonable and why.

6. Repeat these steps with one or two more statements.

These examples show the range of the routine. In the addition example, you should expect to hear that the middle problem (58 + 98) isn't reasonable because 98 is almost 100 and now you're adding many more. The bottom problem (47 + 54) is likely deemed reasonable because both numbers are close to 50 and 50 + 50 = 100. Early work with the routine might call for prompts that are way off (e.g., 9 + 10 is about 100) so that students get a feel for the routine and the reasoning. Though the top examples estimate to 100, you can change them to any value. For example, you might pose that 92 – 57 is about 40 (which is reasonable because 90 – 50 = 40, though 30 would be *more* reasonable).

Is It Reasonable?
37 + 27 is about 100
58 + 98 is about 100
47 + 54 is about 100

Is It Reasonable?
171 – 37 is about 100
182 – 79 is about 100
348 – 250 is about 100

Is It Reasonable?
19 × 7 is about 140
12 × 9 is about 90
23 × 51 is about 100

Is It Reasonable?
73 ÷ 6 is about 20
109 ÷ 9 is about 12
200 ÷ 21 is about 10

ACTIVITY 10.7

Name: "Over/Under"   **Type:** Routine

About the Routine: This routine asks students to compare different problems to a given number, usually a benchmark number (e.g., 50, 100, etc.).

Early work with this routine should pose problems that can be easily determined as over or under the given. For example, it's easy to estimate that 18 + 18 is under (less than) 100. In time, you can begin to mix in problems that will be close to the given. For example, you might ask if 67 + 34 is over or under 100. As these closer estimates are included, you might change the routine to Over/Under/About. So, for 67 + 34, students might say it is about 100 rather than determining whether it will be over or under 100.

Materials: Identify a target number for comparison, prepare two or three addition or subtraction expressions

Directions: 1. Share the over/under number.

2. Pose each expression.

3. Have students estimate the sum or difference of each problem.

4. Students share their ideas with partners, and then discuss their approaches to reasoning with the class.

5. After ideas are shared, the exact solution can be found and compared to the over/under number.

The top expressions are good examples of where you might begin with this routine. Students will determine their estimates in different ways. This is good and it is why discussion is important. On the left, one might say 57 + 48 is over 100 because 57 + 50 is over 100 and that 48 is almost 50. Be prepared for students who share that they don't need to estimate for a given problem. Acknowledge their thoughts and ask them why they don't need to estimate.

> **TEACHING TAKEAWAY**
>
> Introduce estimation with benchmark comparisons. As students' number sense and estimation skills grow, have them estimate and compare numbers such as 75, 30, or 120.

Over/Under 100
57 + 48
35 + 39
69 + 37

Over/Under 50
71 − 29
128 − 43
162 − 83

(Continued)

(*Continued*)

The left example below shows how the routine can be modified for any operation. The right example shows how you can modify the routine for students to consider over, under, or about, as described in the introduction to the routine. The first two problems are likely to be described as over and under, respectively. Some students might think of 24 + 47 as under 75 and some might say it is about 75. This is fine and provides a good opportunity for discussion!

Over/Under 150
6 × 12
9 × 15
20 × 7

Over/Under/About 75
48 + 46
59 + 7
24 + 47

ACTIVITY 10.8

Name: Stay or Go   **Type:** Game

About the Game: *Stay or Go* is a game to practice front-end estimation for addition. Players decide whether their front-end estimate is close (in which case they *stay*) or whether they need to adjust (in which case they *go*). Players try to get the most counters on the game board to win the game. This game can also be adapted to subtraction and to practicing other estimation reasoning strategies. While the instructions are written with two players, this game also works well with two players being on one team and playing against another pair. In this situation, the pairs can strategize what expression to make with their cards to claim a space on the hundred chart (or bump their opponents).

Materials: Bottom-Up Hundred Chart; one deck of cards (queens = 0, aces = 1; remove all tens, jacks, and kings); counters (or markers, if the board is laminated) for marking locations on the hundred chart

Directions: 1. Place deck face down between the partners.

2. Player 1 takes four cards and turns them over side by side to form a pair of two-digit numbers.

3. Player 1 gives the front-end estimate, placing a marker on the appropriate place on the hundred chart.

4. Player 1 decides whether they will stay or go based on the digits in the ones place. Player 2 can challenge this decision (e.g., if the ones digits are 2 and 1, that would not justify going up one more ten).

5. Player 2 repeats these steps to place a counter.

6. If a player's estimate lands on the other player's chip, that chip is returned to its owner.

7. If a player's estimate lands on their own chip, they place a counter on top of it and it is safe (the other player cannot remove it, and if the other player gets that estimate, they cannot place a chip at all).

8. The player who places ten counters first (or who has the most at the end of game time) wins.

ACTIVITY 10.9

Name: Make It Close   **Type:** Game

About the Game: *Make It Close* is a target-based game for practicing estimation of sums or differences. The goal is to create a problem that generates a sum or difference as close to the target as possible. The unique twist in this game is that the target changes from round to round. This dynamic adds value because it offers the opportunity to practice estimating with a wide variety of numbers. When introducing the game, help students think about how to process and estimate. Have them think about what numbers go together to get close to a given sum. If targeting 114, they should think about how 70 and 40 or 60 and 50 both have a sum of about 110. They might even start by thinking 50 and 50 is 100 so they can begin to manipulate their addends.

Materials: Four decks of digit cards (0–9) or playing cards (aces = 1; remove face cards and tens), *Make It Close* recording sheet (optional)

Directions: 1. Players use four digits to make a pair of two-digit addends. The sum of those addends is the target for the first round. Note that the order of the digits creates the addends; the digits are not rearranged.

2. Players then generate four digits to create their own two-digit addends. They can arrange their digits however they like—they can rearrange the digits to get as close to the target as possible.

3. The player closest to the target wins the round. The first player to win three rounds wins the game.

> **TEACHING TAKEAWAY**
>
> Using digit cards instead of dice is helpful for activities that allow students to rearrange digits. The cards help them see the problems they're manipulating.

The upper-left image shows an example of how the game is played. In this round, 6, 3, 5, and 1 were dealt, creating 63 + 51 and a target of 114. Player 1 was dealt 8, 4, 1, and 4. They rearranged the digits to make 44 + 81. Player 2 was dealt 5, 5, 9, and 3. They rearranged to make 53 + 59. Player 2 was closest to the target and won the round.

The example on the left shows how you might modify the game for early work with estimating sums. The lower two examples show how the game could be played with subtraction.

Target	6 3 + 5 1	114	Target	8 4 + 8	92
Player 1	4 4 + 8 1	125	Player 1	7 7 + 6	83
Player 2	5 3 + 5 9	112	Player 2	8 1 + 5	86
Target	7 5 - 2 8	47	Target	2 3 8 - 5 4	184
Player 1	8 3 - 3 1	52	Player 1	2 0 7 - 3 0	177
Player 2	7 1 - 5 5	16	Player 2	1 9 8 - 2 3	175

RESOURCE(S) FOR THIS ACTIVITY

Make It Close

Directions: Use digit cards to make an addition problem. The sum of the problem is the target. Deal new digit cards to make an addition problem that is close to the target. The player closest to the target gets a point.

Target Problem	My Problem
	(give yourself a check mark if you are closest to the target)

online resources — This resource can be downloaded at **https://qrs.ly/psf6a5o**

ACTIVITY 10.10

Name: High/Low Estimates   **Type:** Game

About the Game: This is a twist on High/Low Doubles (Activity 9.6) in which players estimate sums or differences. In this version, one player is the predictor trying to get six predictions in a row while the other player is the dealer and the checker. In the game, the player predicts if a card will be more than or less than a given benchmark (e.g., 100). If they're correct, they keep the card. If not, they lose. The goal is to get six predictions correct in a row.

Materials: Index cards with comparison problems (see the chart below), a calculator

Directions: 1. Note that before the game is played, it must be clear to students that half of the cards are more than the given benchmark and half are less than the benchmark.

2. The player predicts if the first card's answer will be more than or less than the benchmark.

3. The card is shown and the player states whether their prediction is correct and how they know.

4. The dealer finds the sum to see if their estimate/prediction is correct. Note that using a calculator quickens play and ensures accuracy.

5. If the player is correct, they take the card. The process is repeated with the player predicting and estimating before the dealer confirms.

6. The goal is for the player to get six cards before they bust. If they bust, the dealer wins. Players then reverse rolls.

The table shows problem cards you can use for different versions of the game. You can write these on index cards, put them in a Ziploc bag, and label it with the version (e.g., Benchmark Comparison 100).

BENCHMARK COMPARISON 100		BENCHMARK COMPARISON 50	
Less-than cards	More-than cards	Less-than cards	More-than cards
59 + 18 =	59 + 78 =	29 − 12 =	112 − 24 =
35 + 63 =	63 + 42 =	68 − 59 =	68 − 12 =
52 + 42 =	77 + 39 =	71 − 45 =	91 − 15 =
27 + 49 =	47 + 69 =	86 − 57 =	126 − 57 =
67 + 23 =	67 + 83 =	93 − 82 =	140 − 22 =
18 + 15 =	78 + 71 =	55 − 12 =	150 − 78 =
45 + 50 =	69 + 53 =	85 − 63 =	68 − 13 =
25 + 34 =	25 + 94 =	60 − 17 =	81 − 17 =
15 + 75 =	31 + 87 =	91 − 67 =	109 − 45 =
27 + 46 =	51 + 66 =	62 − 41 =	112 − 57 =

BENCHMARK COMPARISON 1,000		BENCHMARK COMPARISON 500	
Less-than cards	More-than cards	Less-than cards	More-than cards
185 + 115 =	785 + 715 =	550 − 125 =	1,550 − 786 =
450 + 505 =	690 + 535 =	658 − 639 =	688 − 139 =
258 + 347 =	258 + 947 =	606 − 597 =	816 − 197 =
315 + 575 =	315 + 875 =	690 − 425 =	990 − 125 =
207 + 465 =	517 + 665 =	621 − 571 =	1,121 − 570 =
593 + 218 =	595 + 782 =	716 − 535 =	950 − 125 =
235 + 263 =	633 + 426 =	1,108 − 999 =	1,228 − 199 =
452 + 442 =	477 + 839 =	907 − 709 =	1,606 − 1,097 =
127 + 749 =	547 + 569 =	819 − 475 =	1,140 − 375 =
567 + 223 =	967 + 183 =	922 − 563 =	821 − 171 =

ACTIVITY 10.11

Name: Chips   **Type:** Game

About the Game: *Chips* is a two-player game where players try to capture their opponent's five chips. Each chip has a benchmark number written on it (e.g., 50). Players take turns generating a problem and estimating its sum or difference. If their estimate matches an opponent's chip, they take the chip. There are two different ways to play the game.

First, you can have students use dice to generate problems. A student might roll four dice to get a 7, 3, 3, and 8 (or pull four cards with a 7, 3, 3, and 8). They could make the problem 73 + 38, 38 + 37, 33 + 87, and so on.

Second, you can make a collection of cards with problems written on them (similar to Activity 10.10). Players take turns flipping cards and estimating the results. Possible problems to use with cards are listed in the table below.

Materials: For a single pair of students, you'll need twenty counters or plastic chips with target numbers written on them (see the chart below) and four decks of digit cards (0–9), playing cards (aces = 1; remove face cards and tens), or four 10-sided dice

Directions: 1. Each player takes five chips from a bag and puts them out so their opponent can see them.

2. Players take turns generating a problem and estimating its sum or difference.

3. If an estimate matches an opponent's chip, the player takes the chip.

4. The first player to take all of their opponent's chips wins.

For example, Player 2 pulls out the five chips shown below. On Player 1's first turn, they make the problem 39 + 38. Player 1 says, 39 + 38 is about 80 because 40 + 40 is 80. Player 1 takes Player 2's chip (80). Then, it's Player 2's turn to make a problem.

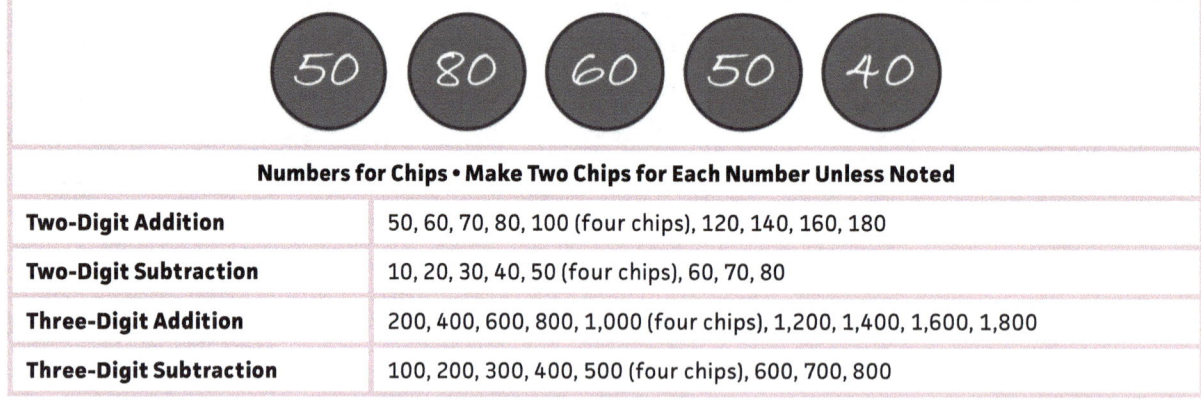

Numbers for Chips • Make Two Chips for Each Number Unless Noted	
Two-Digit Addition	50, 60, 70, 80, 100 (four chips), 120, 140, 160, 180
Two-Digit Subtraction	10, 20, 30, 40, 50 (four chips), 60, 70, 80
Three-Digit Addition	200, 400, 600, 800, 1,000 (four chips), 1,200, 1,400, 1,600, 1,800
Three-Digit Subtraction	100, 200, 300, 400, 500 (four chips), 600, 700, 800

Using cards is a fine way to play the game, too. The cards here are examples from the online site that can be edited as needed. Problems could also be recorded on index cards.

CHIPS 2-digit Addition		
38 + 39	43 + 41	59 + 63
49 + 53	18 + 41	31 + 57
29 + 46	25 + 34	31 + 18
81 + 74	83 + 17	55 + 25

CHIPS 2-digit Subtraction		
55 − 36	86 − 24	60 − 17
99 − 15	102 − 15	78 − 57
65 − 25	108 − 74	68 − 45
55 − 29	48 − 29	97 − 53

CHIPS 3-digit Subtraction		
335 + 463	524 + 142	227 + 449
557 + 523	591 + 188	529 + 467
249 + 672	135 + 465	648 + 155
581 + 903	216 + 933	369 + 451

CHIPS 3-digit Subtraction		
294 − 128	682 − 559	710 − 535
286 − 207	329 − 180	357 − 217
532 − 219	1,057 − 617	527 − 184
578 − 497	783 − 747	1,225 − 568

 This resource can be downloaded at **https://qrs.ly/psf6a5o**

ACTIVITY 10.12

Name: Make Five Estimates

Type: Center

About the Center: Make Five Estimates is a center for students to practice estimating sums. Students make problems that have estimated sums that match conditions on their cards. The goal of the center is to get all five estimate cards in the fewest number of attempts. The recording sheet asks students to show how they know their problem works. This is where you can see their thinking. It will help you determine if students are truly estimating or if they are finding exact sums and then creating an estimate. The online resource provides cards for addition and subtraction of two- and three-digit numbers.

Materials: Playing cards (queens = 0, aces = 1; remove tens, kings, and jacks) or four sets of digit cards, one set of Make Five Estimates cards, Make Five Estimates recording sheet

Directions:

1. Students shuffle the cards and stack them in a deck face down.

2. The first card is flipped over (e.g., Sum Is More Than 100).

3. The player makes a problem with a pair of two-digit addends and estimates the sum.

4. If the sum matches the card, the player records the card, their problem, and how they know the estimated sum matches the card.

5. If the estimated sum doesn't match, the student makes a tally at the bottom of the recording sheet for their first attempt and makes a new problem. If that problem matches, they record it; if not they add another tally and try again.

6. The student tries to match five problems to five estimate cards in the fewest number of problems.

> **TEACHING TAKEAWAY**
> You can change the number of estimates needed from five to three if you have a limited amount of center time during your class. You could also have students continue a center over multiple days.

In this example, a student was using Make Five Estimates cards for two-digit addends. The first card they flipped over said, "Sum Is About 100." The first problem they made was 72 + 97, which isn't about 100. The player made a tally (bottom image) and then tried a new problem. The second problem was 84 + 68, which doesn't match the card either and a second tally was added. The third problem was 41 + 65. The student said it did match and explained their thinking that 40 + 60 is 100 and that 41 is close to 40 and 65 is close to 60 (top image).

My Card	My Problem	My estimate matches because
Sum is about 100.	41 + 65	40 + 60 is 100 and 41 is close to 40 and 65 is close 60.

Use tally marks to keep track of the number of problems you had to make to get all 5 estimate cards. ||||

RESOURCE(S) FOR THIS ACTIVITY

Make Five Estimates

Directions: Deal five estimate cards. Create a problem and estimate the solution. If the estimate matches one of your cards, record it on your recording sheet and explain why it fits. Try to make all five estimates in the fewest number of problems.

My Card	My Problem	My Estimate Matches Because

Use tally marks to keep track of the number of problems you had to make to get all five estimate cards.

This resource can be downloaded at **https://qrs.ly/psf6a5o**

ACTIVITY 10.13

Name: *Sorting Estimates*   **Type:** *Center*

About the Center: In this center, students sort problems based on their estimation of the result. After they sort a card, have them use a calculator to find the exact answer to confirm their estimation.

To maximize the effect of practice with this routine, have students select one problem from each category after they have sorted all of the problems. For each problem they choose, have them explain their estimate and compare it to the actual answer.

Materials: Problem cards, recording sheet (optional)

Directions:
1. Give students a sorting mat (optional) or have them create three columns in their mathematics journal—a column for *more than*, *less than*, and *about*.

2. Give students a dozen or so problems and a number for comparing their estimates to.

3. Students select a problem, estimate the result, and sort the estimate in comparison to the benchmark number you assigned.

4. After sorting the problem, students find the exact answer with a calculator. (Note: See the Marginal Teaching Takeaway for a calculator alternative.)

5. If the estimate is sorted appropriately, they keep the problem on their chart. If not, they remove the problem from the collection. This is optional, but it allows you to assess how well they're estimating. Alternatively, you can have them record problems estimated correctly and incorrectly in a two-column chart in their journal.

6. After students have sorted all of the problems, they choose one problem from each category, explaining how they recorded their estimate and how the estimate compared to the actual result.

Here are examples of the center for addition and multiplication. You can see how the students sorted their problems. This teacher decided to label the cards and provide an answer key for students to check their work. The teacher can easily change the target number, the numbers in the problems, or the operation to use the center over and over.

> **TEACHING TAKEAWAY**
>
> Using a calculator is a good way to help students check their work. Alternatively, you can label each problem card with a letter and make an answer key for students to check their work after sorting.

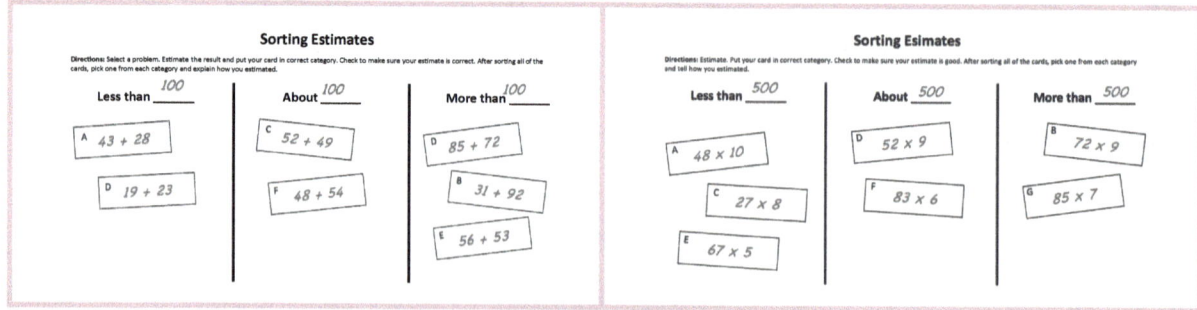

A blank recording sheet can be downloaded at **https://qrs.ly/psf6a5o**

ACTIVITY 10.14

Name: Just One, Which One   **Type:** Center

About the Center: This center is an independent practice to reinforce the notion of estimating a solution by adjusting one number in a problem (see Activity 10.4). It's a concept that students will need to practice, especially if their estimation experiences have been mostly procedural or focused on finding two benchmark numbers.

Materials: 10-sided dice, playing cards (queens = 0, aces = 1; remove tens, kings, and jacks), or digit cards; recording sheet (optional)

Directions:

1. Students generate a problem with dice or cards. Alternatively, you can provide problem cards written on index cards or printed on paper.

2. Students record the problem.

3. Students estimate the result by changing one number to a benchmark number and record the estimate.

4. Students estimate the result a second time by changing the second number.

5. Students identify which problem was easier for them to think about.

The recording sheet is shown. Students could fold a piece of paper in thirds, draw columns on a dry erase board, or write in their mathematics journals. Notice the extension added at the bottom of the recording sheet, asking them to explain whether it's easier to estimate by adjusting one or both numbers. You can swap that out with other prompts such as, "Tell me which problems were easier to estimate and why" or "Were there any problems that were easy for you to estimate if you changed either number? Which problems were they? What made them easier?"

Just One, Which One

Directions: Generate a problem and record it in the first column. Estimate the result by changing one number. Then estimate again by changing the other number. Star the one that's easier to think about.

Problem	Estimate #1	Estimate #2
77 + 57	80 + 57 = 137 ★	77 + 60 = 137

What are you noticing about which number to change to estimate?

 This resource can be downloaded at **https://qrs.ly/psf6a5o**

NOTES

PART 3

TAKE ACTION

It is time to take action, putting your learning in motion to help your students strengthen foundations for fluency—in other words, creating a plan to provide good and necessary beginnings. Determine what works best for your students and your instruction. Identify their needs. Prioritize the activities and resources that meet those needs. It may take some tinkering and some trial and error, but you cannot give up on the quest to develop every student's fluency. You are not only developing their concepts and skills but also their positive mathematics identities and agency.

Through the modules, you have become acquainted with instructional activities, routines, games, and centers to use in your classroom to teach as part of first instruction or to shore topics as part of intervention. As you think about what to do to build strong fluency foundations, some *how* and *when* questions are probably surfacing. There isn't an exact recipe to follow in any case. Every situation is unique from students' strengths to time allotments for instruction to core instructional resources to the human resources that can help with the work. Thus, Part 3 is designed in a question-and-answer format. To begin, ask yourself:

- What are my/our computational fluency goals?

- What foundations will my students need in order to be successful with the broader computational fluency goals?

- What will I/we need to do to be successful?

With these broad questions in mind, begin to put together a game plan that you anticipate will meet your students' needs. Read the sections that will help you refine and implement your plan. Revisit your plan and reread relevant sections of Part 3. Finally, as we say at the end of this part, celebrate your successes as you journey to figure out fluency in your setting!

TAKING ACTION IN THE CLASSROOM

As with any new action, many dilemmas arise and thus decision-making and reflecting on those decisions is essential. One dilemma in taking action in your classroom is … well, time. When might you work the instructional and practice activities into your teaching block or day? Another time dilemma is deciding which topics are necessary for students' fluency development and how much time you want to devote to each topic (and thus, what assessment tools will be used to determine priorities). Beyond time commitment, there is the question of whether you are focusing on first instruction or review. Finally, how will you monitor progress to know that your efforts are working?

As you implement your plan, take some time and reflect on how each step went. Reflect individually, then talk with teammates and colleagues. Reflect not only on what isn't working but also on what is working and why. Ask yourself the following questions:

- What am I learning about my students' strengths and needs?

- What has been most successful? Why is this the case?

- What should I keep doing?

- What isn't going as planned? What might be contributing to this?

- What adjustments need to be made for the future?

Speed bumps with this work are to be expected. Before overhauling your entire plan, consider slight, intentional adjustments over time. But before you can adjust a plan, you have to make it! The following questions are posed to help you think about what to consider and to ultimately help you form your plan.

IS THIS WORK PART OF FIRST INSTRUCTION, REVIEW, OR INTERVENTION?

This is a critically important first question. The answer will color many of your actions. Essentially, you are asking if the skills and concepts in this book are being used to complement your first instruction of grade-level standards. If so, you might use activities from one module as a mini unit. Are you using them as a primer before you embark on teaching fluency strategies? If so, you might space this work out across the weeks leading up to the fluency topic. You might decide to use all of the resources but rearrange them to line up with your scope and sequence. You might reserve some of the resources for reteaching a small group along the way. You might limit the size of the numbers you work with relative to your students' mathematical maturity and experience.

Are you providing a review of skills that were previously taught but that students haven't reached automaticity or fluency with yet? In other words, are you a third-grade teacher who needs to provide more experiences with skip counting or decomposition because your students are inefficient when adding 193 + 340? If this is the case, you might be targeting a specific foundation. You might need to use all of the instructional tasks, prompts, and practices. You might use them first with two-digit numbers then build toward three-digit numbers and so on.

WHAT FOUNDATIONS SHOULD I FOCUS ON FIRST?

In short, the answer is, "It depends." One thing it depends on is your grade-level content. In terms of first instruction, arrange topics that align with your scope and sequence. If you are about to be teaching two-digit subtraction, then you will want to explore some of the activities on decomposing (Module 2), skip counting (Module 4), and properties of addition and its inverse relationship with subtraction (Module 5).

A second consideration is which foundations students most need to develop. To begin, check on what your students learned in the previous grade. Don't assume that they are strong with any prerequisite skill but also don't feel compelled to create some sort of prerequisite assessment for each unit of study. Instead, use one of the quick assessments in these modules or implement a routine and see how it goes. These are both short time investments that give quick data on which modules might need to be reviewed first. Gather information about your students' current understandings. As students work with tasks in your lesson, take note of how well they use the foundations in this book. Pay attention to how they skip count, estimate, or break apart numbers. As they are working, stop and ask individual students, "How are you thinking about this problem?" Keep track of what methods you notice your students using and those you don't see very often. For example, do they always skip count forward by groups (e.g., by tens, fifties, or hundreds) yet always skip count backward by singles (e.g., ones and possibly tens). Do they skip count by tens well within a century but stop to count by singles around a century? If you see your students are not using estimation skills, provide opportunities for them to do so. Do the same for relating addition and subtraction along with other ideas that are foundational for fluency.

This information helps you determine your students' greatest needs. Begin with those. And remember, regardless of your purpose—first instruction or continued experiences with previously learned content—you don't have to work through the modules as they are presented in the book.

WHEN DO I USE THESE ACTIVITIES?

There are different opportunities throughout the day when you might use an activity. While it is important to ensure you are teaching your school or district's curriculum, these activities serve as supplemental lessons and practice opportunities to complement your curriculum. This is especially helpful when your students' understanding or skill with a topic isn't fully realized through a single lesson on the topic followed by a smattering of practice in a unit.

DO I IMPLEMENT THESE ACTIVITIES DURING MY MATHEMATICS LESSON BLOCK?

Yes, they can be done during mathematics as part of your first instruction, either as a full lesson or a warm-up or closing activity. You can insert an instructional activity into the series of lessons within a unit to reteach or go deeper with a topic. For example, a second-grade teacher might find a few skip-counting lessons within an addition unit. That teacher might add Activity 4.2 as another lesson. A given lesson might be done in a whole-group setting, a small-group rotational setting, or with a subset of students who need more experience with the topic.

The routines, games, and centers in this book naturally complement what you already do. You can simply pull an activity from this resource and use it within your daily structure. For example, if you notice that your students need more practice with subitizing, you might swap your opening routine for Activity 2.6 or Activity 2.7 for a time. Another option could be to use a game or an activity such as Activity 2.11 as an independent practice.

DO I IMPLEMENT THESE ACTIVITIES OUTSIDE OF MY MATHEMATICS BLOCK?

Yes, it makes sense to use these resources outside of mathematics class, too. These foundations are essential and students will benefit from extended opportunities to learn and practice these foundations. The activities are good practice opportunities for students as they arrive in the morning, come in from lunch, or during the time between packing up and dismissal. Taking advantage of different parts of the day for a little extra practice is a great idea. A fluency activity could be sent home with students as a homework assignment. Fluency games that students have previously played in the classroom may be sent home during extended breaks or days off to encourage students to practice. Any and all opportunities for practice will contribute toward strengthening students' foundational fluency skills.

CAN THE FOUNDATION ACTIVITIES BE USED IN SMALL GROUPS?

Foundation activities can definitely be used with small groups of students. There are times when all students in your classroom don't need reteaching or practice with the same foundation or when you may want to focus your attention on a particular group of students to assess how well they understand and use certain skills. By putting them in a small group and working with them directly, you can better observe and assess as well as ask follow-up questions to probe their thinking more deeply. These scenarios are perfect for utilizing a small-group structure for fluency activities.

One approach involves using a small-group rotation model. In this model, students are together for the opening routine and then split into two groups. One group works with the teacher while the other works independently and then the groups switch. This model allows every student to receive small-group instruction. The groups may work with the same content or a slightly different content, based on the needs of each group.

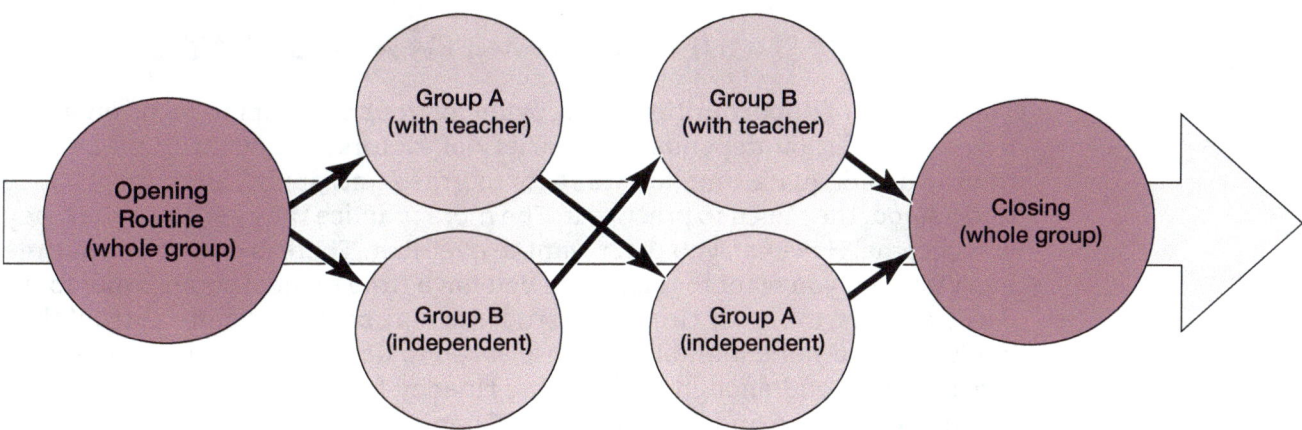

Another approach is a small-group breakout model. This option begins with teaching a skill or concept with the whole group before breaking into smaller groups for the latter part of the lesson. One group will meet with you for more work with the skill or concept of the lesson. The other group will practice the skill or concept independently. For some lessons, you might choose to extend or enrich the concept with the students in your group rather than always focusing on reteaching or reinforcing.

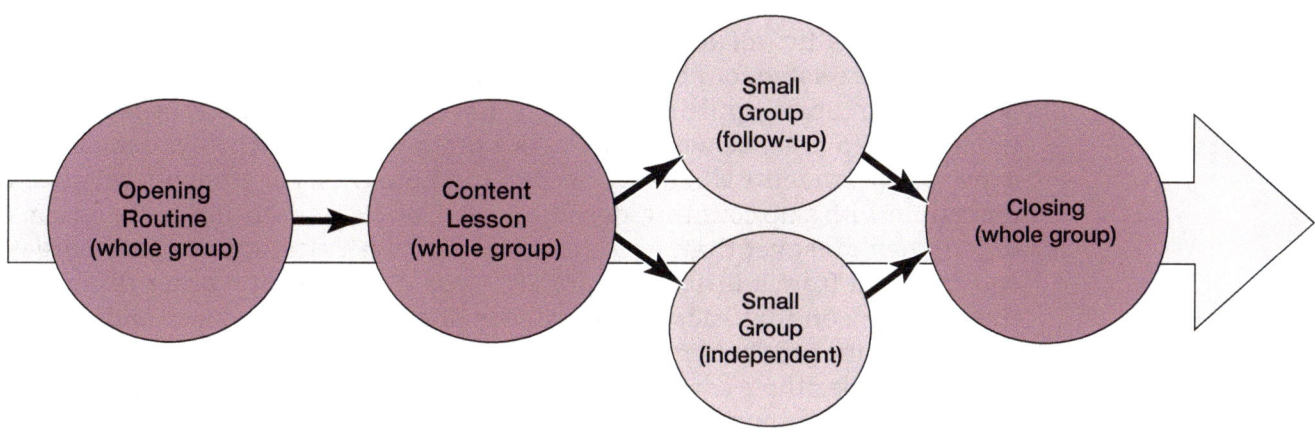

HOW DO I MAKE TIME FOR THIS? WHAT SHOULD I LET GO OF TO MAKE TIME FOR FLUENCY?

You might be thinking, "I want to do this, but I don't have time as it is!" Consider replacing old routines with new ones that build fluency foundations. Instead of having students work through stacks of flash cards when they return from recess or work on an online fluency practice program (really a string of problems to solve), have them engage in a more interactive fluency activity instead. Take a closer look at the games in your independent stations. Consider the skills that those games practice and think about what skills your students are competent at. Are those games necessary? Think about free time or indoor recess. Games are a nice option for students to use during those moments. In short, you know that fluency and the foundations in this book are a priority for you and a need for your students. You can find time to make it happen by letting go of something else. When students are fluent with these foundations, then other content comes more readily. So, invest early in the foundations and you will get time back when teaching the operations.

HOW LONG SHOULD I STAY WITH A FOUNDATION?

The number of opportunities you provide for your students to focus on a particular foundation depends on a variety of factors. You want to make sure to provide adequate time for students to grasp and apply the concept. Once understood, they need to practice it. The more practice they get, the better they will become. However, you don't want to overdo it. There is only so much time in a year, and you want to make sure you have time to incorporate other foundational ideas, too. Over time, your students may become disinterested if they work with the same foundation or are asked to do the same type of activities repeatedly. Spend enough time on one fluency foundation for your students to gain confidence and become better. Then move on to another topic. After moving on, go back and revisit the topic from time to time! Coming back to a foundation is especially useful when introducing a new topic that makes use of a foundation from earlier in the year.

WHAT IF NOT ALL OF MY STUDENTS NEED TO PRACTICE?

In literacy, students often read different books because they have different strengths and needs. There are times when everyone experiences the same thing, especially when something new is introduced. But then, students break off for differentiated instruction and practice to target student needs. There isn't a one-size-fits-all approach. The same idea applies to computational fluency and the related foundations. Some students will need extensive practice with certain topics, while others might have great utility with these foundations but need practice with others. As a mathematics teacher, you must find ways to give students what they need in mathematics. For students who need more time with skip counting, decomposition, or distance to and combinations of ten, they must get that. And though others with strengths in these areas could benefit from a little more practice, you'll need to find other things for them to work on that address their needs or extend their understanding. For example, some students might need more time with decomposing two-digit numbers while others seem to be in a good place. The latter group could dabble with decomposing numbers into three parts (e.g., breaking 96 into 75, 20, and 1).

You might let them play with ideas of manipulating numbers through subtraction (e.g., 96 could be thought of as 100 and -4) or they could begin to decompose three-digit numbers. Create games and centers from across the modules so that you can pick and choose the ones that are the best fit for each student.

HOW DO I KNOW EXACTLY WHICH FOUNDATIONS TO WORK ON?

As you consider your students, you are probably wondering which exact foundations you should focus on. Unfortunately, there isn't a magic recipe. Be strategic by gathering evidence of what your students know and which foundations aren't fully developed quite yet. Use this information to make a plan of attack that best meets your students' current needs. Continue to monitor student progress to make the adjustments to the plan as you go.

A good place to start is by paying close attention to what students do in class and especially how they compute (assuming your students have learned fluency strategies). Check to see how well they break numbers apart, count and skip count, make tens, and so on. Consider if there are patterns in their thinking (e.g., do they always count on by groups and count back by singles?). Observing and noting strengths and needs during instruction can help you identify the foundational concepts or skills you need to revisit.

Once you target a foundation, turn to that module in this book. Use one or two of the quick assessment ideas to get more information about what the student(s) knows or create a checklist of subskills to monitor or look for using the look-fors in the assessing section of each module. In the following example, you see a teacher who created a checklist for four students who she noticed seemed challenged by breaking apart numbers. She created her checklist from the information in Module 2.

	AIDEN	ASIA	BRAXTON	JULIA
Subitizes small numbers	√	√	√	√
Subitizes larger numbers by seeing parts	√	√	√	√
Decomposes by place value		√		√
Decomposes a number in a variety of ways		√		

TAKING ACTION BEYOND THE CLASSROOM

It takes a village to help a student realize their fluency. The first few pages were considerations for you (the teacher) to provide first instruction or review in the regular classroom. Many students (including those not in intervention settings) need lots of practice with fluency foundations beyond the designated mathematics block. Practice at home with a caregiver can help provide ongoing enjoyable experiences to strengthen students' skills (e.g., sending a game home and having students play it at least twice a week). Some students need more specialized and personalized fluency support from an interventionist, special educator, or possibly a private tutor. In each case, the adult who's working with your students can be a great asset.

CAN AN INTERVENTIONIST USE THESE ACTIVITIES AND RESOURCES?

Absolutely! Practice in intervention settings, including online apps, can often be relegated to recalling a memorized procedure and completing that procedure over and over. This type of intervention does not attend students' missing or fragile concepts and skills. Evidence-based instructional practices that support students in intervention settings include (Fuchs et al., 2021):

- explicit instruction that focuses on developing student understanding of mathematical ideas,

- supporting students' use of the language to help them effectively communicate their understanding of mathematical concepts,

- using well-chosen concrete and semi-concrete representations to support the learning of concepts and procedures (and the connections among them), and

- using the number line to support the learning of concepts and procedures.

Note the importance of conceptual understanding and representations to support student learning! Activities in these modules reflect these evidence-based strategies. When an activity does not have a representation, it can readily be added by, for example, inviting students to illustrate their thinking on an open number line. Each module's reference page offers suggestions for representations and highlights key ideas. Thus, these modules are an excellent fit for intervention, providing students with good beginnings so they can have access to the grade-level concepts and procedures they are learning!

The activities in this book can be used with individuals or small groups of students. They may become the focus of a daily intervention lesson for students who are still in need of acquiring (learning) the skill. Games and centers are perfect intervention tools for students who show understanding but need more repetition. Games and centers provide more engaging and meaningful opportunities for students because they use visuals, think-alouds, and peer support and are low-stress and motivating. In other words, games and centers are not only more enjoyable, but they are also more effective in supporting students who struggle or who have special needs in mathematics. Even with a prescribed intervention curriculum or program, the activities in this resource are good options for complementary work that can be used in addition to the program or in place of some component of it.

ARE ACTIVITIES APPROPRIATE FOR STUDENTS WITH SPECIAL NEEDS?

Most definitely! Students who receive special education services often need additional time beyond what is provided in the classroom to develop understandings of these foundational ideas for fluency. Focusing on the ideas presented in this book will be more beneficial than providing drills and practice that rely on memorization, which is typically not a strength of students with special needs. Students with special needs benefit from the research-based strategies listed above. The foundations in this book are often the very things they need more experience with. Use these topics to shape their education plans. And be sure that explicit strategy instruction doesn't become a call for mimicry but rather a request for reasoning.

HOW MIGHT I INVOLVE STUDENTS' CAREGIVERS IN DEVELOPING STRONG FOUNDATIONS?

Utility with these skills comes through understanding and practice. Teaching them happens in your classroom or intervention group. Practice, while certainly part of your instructional model, can go beyond the school day to include experiences at home. Most often, caregivers are more than happy to help support their child's learning at home, though they may have anxiety about doing mathematics and will benefit from enjoyable practice and guidance from you on how to support the foundations. Take time to communicate to caregivers why fluency practice matters and how these foundations contribute to it. One way to help caregivers understand strategies and foundations is to provide a visual-rich, brief explanation of the foundational concept or skill; for example, see the Foundation 1: Number Relationships reference page illustrated here:

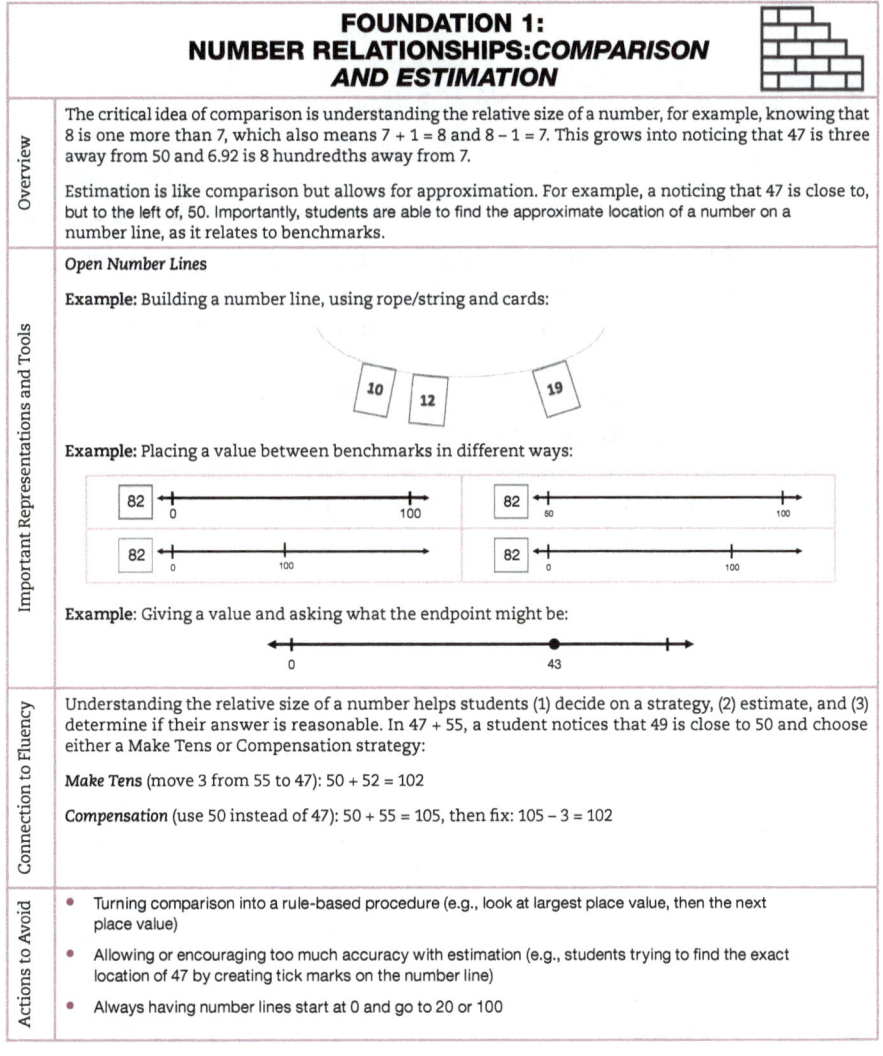

FOUNDATION 1: NUMBER RELATIONSHIPS: COMPARISON AND ESTIMATION

Overview

The critical idea of comparison is understanding the relative size of a number, for example, knowing that 8 is one more than 7, which also means 7 + 1 = 8 and 8 − 1 = 7. This grows into noticing that 47 is three away from 50 and 6.92 is 8 hundredths away from 7.

Estimation is like comparison but allows for approximation. For example, a noticing that 47 is close to, but to the left of, 50. Importantly, students are able to find the approximate location of a number on a number line, as it relates to benchmarks.

Important Representations and Tools

Open Number Lines

Example: Building a number line, using rope/string and cards:

Example: Placing a value between benchmarks in different ways:

Example: Giving a value and asking what the endpoint might be:

Connection to Fluency

Understanding the relative size of a number helps students (1) decide on a strategy, (2) estimate, and (3) determine if their answer is reasonable. In 47 + 55, a student notices that 49 is close to 50 and choose either a Make Tens or Compensation strategy:

Make Tens (move 3 from 55 to 47): 50 + 52 = 102

Compensation (use 50 instead of 47): 50 + 55 = 105, then fix: 105 − 3 = 102

Actions to Avoid

- Turning comparison into a rule-based procedure (e.g., look at largest place value, then the next place value)
- Allowing or encouraging too much accuracy with estimation (e.g., students trying to find the exact location of 47 by creating tick marks on the number line)
- Always having number lines start at 0 and go to 20 or 100

Foundation summaries can be downloaded for each module at **https://qrs.ly/psf6a5o**

Downloads can be readily adapted to complement the textbook you are using or to include student work from your classroom. Even shorter resources can be created and inserted into newsletters or shared at family events (e.g., available at a table where they have played a game related to that foundation).

Remember, it's highly likely that caregivers did not learn fluency foundations or reasoning strategies in school. As adults, they likely already skip count, decompose, and make tens well. They may have a different idea of what practice (especially mathematics practice) is than you do. They may think rote practice is the goal. So, give advice on what to ask or what to look for when working with the student. Each module offers important look-fors that you can share in a newsletter or a unit overview. Provide questioning prompts to caregivers so they have examples of how to invite their child to explain their thinking. Use family events to highlight the importance of reasoning, sensemaking, and learning the foundations. Share a game with families that can be played many times. Celebrate and appreciate their efforts with a quick note of thanks by email or a call home.

MONITORING STUDENT PROGRESS:
What Evidence Should I Collect and How Do I Evaluate It?

There are different ways to collect data about your students' growing skillset. A brief paper-and-pencil assessment is familiar and fine. It can give you a good artifact to corroborate what you see in class. As you have noticed, there are quick assessments at the beginning of each module. The instructional prompts within each module—such as the two below from Module 3—are additional options that you can use to gather that evidence:

- Two different numbers make 100. What are three possible sets of numbers? Choose one of your sets and prove that it does make 100.

- Luka says that because $2 + 8 = 10$, he knows that $20 + 80 = 100$ and $200 + 800 = 1,000$. Is Luka's thinking on track? Share why you think it is or isn't.

Observation is a powerful way to gather evidence. Unfortunately, this evidence can be discounted because it doesn't look like a traditional assignment, quiz, or commercial screener. Games and centers along with observation during lessons are a great way to monitor progress. You can also observe students employing skills and strategies as they work through mathematics class. Use and document your observation! To do this, create a checklist with foundational fluency concepts, as shown on page 227. When you observe a student using or applying the skill with ease, put a check by their name to indicate you saw evidence of it in action.

Using a rubric to evaluate student understanding is a good way to monitor progress as you teach or reteach a foundation. The rubric below is one of many options to do this, as it is good for examining student thinking, reasoning, justification, representation, and so on.

Not Yet Demonstrating Understanding	Demonstrates Some Understanding		Demonstrates Understanding
Answer/Solution Is Incorrect	Answer/Solution Is Correct	Answer/Solution Is Incorrect	Answer/Solution Is Correct
❑ Justification is mathematically incorrect, missing, or irrelevant. ❑ Representation is mathematically incorrect, missing, or irrelevant. ❑ Strategy is mathematically incorrect, missing, or irrelevant.	❑ Justification is incomplete, slightly flawed, or unique to the task. ❑ Incorrect answer is the result of imprecision, though the justification is accurate. ❑ Representation is relevant but mathematically flawed. ❑ Strategy is valid but inefficient, inappropriate, or unique to the task.		❑ Student enacts the foundation (skill) with efficiency and accuracy. ❑ Student carries out the foundation (skill) with automaticity, needing little thought, effort, or cognitive demand. ❑ Student is able to use the foundation (skill) in many, if not all, situations.

This resource can be downloaded at **https://qrs.ly/psf6a5o**

Be mindful that the goal with these skills and concepts is automaticity. That is, you want students to use and apply the skills without having to think about it a lot. The point is that a rubric that monitors and evaluates progress of understanding or attaining the concept (similar to the one above) doesn't assess the end goal—applying the concept or skill effortlessly. Because of this, you might look to assess competency with the foundations using a rubric similar to this one:

Not Yet Demonstrating Understanding	Demonstrates Understanding	Demonstrates Competency
❑ Student is unable to carry out the skill. ❑ Student is unable to explain a strategy for using the skill.	❑ Student is able to carry out the skill correctly but does so inefficiently. ❑ Student is able to explain how they carried out the skill. ❑ Student is able to use skill in some situations but not all.	❑ Student completes the skill with efficiency and accuracy. ❑ Student carries out skill with little thought, effort, or cognitive demand. ❑ Student is able to use skill in many, if not all, situations.

This resource can be downloaded at **https://qrs.ly/psf6a5o**

MONITORING STUDENT IDENTITY AND DISPOSITION

Mathematics identity—the way students see themselves as learners or doers of mathematics—impacts their learning (Aguirre et al., 2013). Progress with learning mathematics content must be monitored, but also consider monitoring the emergence of positive mathematics identities among your students. Additionally, how students feel about mathematics (Do they like it? Do they think it is important?) is important to monitor. This is extremely important for students who are learning in intervention settings, who may discount their ability to do and use mathematics because of their setbacks, especially when their classmates seem to do fine with the very same content and experiences. In other words, they can have a negative mathematics identity. Thus, they don't feel that putting in effort will lead to a positive outcome (i.e., a correct answer or good estimation). Attending to identity, dispositions, and content simultaneously can reap benefits for all three.

Activities to help you learn more about your students' dispositions don't need to take long. You might utilize a survey in which students draw either a smiley face or a sad face to answer questions about how they feel about mathematics. Another survey may be as simple as asking questions about how they perceive themselves as mathematicians to the class, with students responding with a thumbs up or thumbs down. Statements you might use include the following:

- I can do mathematics.

- I like mathematics.

- Mathematics is important.

- I am getting better at _____ (insert concept/skill).

- I can figure out a way to do problems when I'm unsure.

Other activities to get to know who your students are could include a questionnaire in which students list the top three things they are good at doing in mathematics and the two things they feel they need to work on. Another identity activity would be to ask your students to draw an emoji to describe how they see themselves as mathematicians (SanGiovanni et al., 2020). These "getting acquainted" activities shouldn't take up the valuable time you could be using to build fluency. However, allocating a few minutes here and there will give you a better idea of who your students are as learners of mathematics. They can be part of your independent time in mathematics class, as part of your closure (from time to time), or as part of a homework assignment.

HOW DO I DETERMINE WHETHER MY STUDENTS ARE SUCCESSFUL?

These foundations are needed for life, so monitoring success can occur throughout the year. For specific guidance on measures of success, check out these contents in each module: look-fors (a checklist of what knowing that foundation looks like), quick assessments (six possible prompts to choose from to assess student understanding and skill), and the prompts for each activity (usually the fifth activity in the module, which includes a variety of prompts for written responses or interviews). You can also observe success in everyday interactions with students as you will see them in action! You'll see that skill with skip counting is improving because a student will count by hundreds to solve 948 − 613. You'll see flexible decomposition when students compensate to solve 63 − 48. You'll hear them question their answers by talking about reasonableness.

Some students will catch on quickly and be able to apply these foundations effortlessly. Does this mean they should not participate in these activities anymore? Not necessarily. Students who show skill in one setting might not be fully comfortable using the skills in other situations or with different tasks or concepts. These students need to revisit the foundation every so often to maintain their understanding and continue to develop their automaticity. You might think of it as similar to playing the piano: If someone takes a long break from playing or practicing, they may be clunky when they start up again. But with more routine practice, this regression doesn't seem to happen. The same is true for these foundations. Intermittent practice will help students maintain and improve their competency with the concepts and skills. Keep in mind that you can provide targeted practice in isolation, but foundations are also practiced as students apply them to other topics, such as developing their fluency with addition or multiplication of whole numbers.

Some students will be further along than others. Understandably, you (and they) might become a bit frustrated when progress stalls or isn't uniform. Remember, all students develop at different rates and at different times. Be patient, as some students will need more time than others to become adept with these foundations.

It is important to celebrate progress—no matter how big or how small—and to remain committed to the goal. Your students can (and will) figure out their fluency, especially when foundational understandings and skills shift from challenges to assets. Celebrate their growth in reasoning and share your successes with your colleagues, families, and professional community. Figuring out fluency is a journey, and it is important to acknowledge and highlight successes, lessons learned, and progress made!

Appendix
Tables of Activities

This book is packed with activities for instruction, practice, and assessment to support the work that you do and to supplement the resources you use. The following pages provide a listing of all of the activities in this book. These tables can help you achieve the following:

- Jump between foundations, as you may not teach them sequentially.

- Locate prompts for teaching each foundation.

- Identify a specific type of activity to incorporate into your instruction.

- Identify activities for specific foundations that you need to reteach, reinforce, or assess.

- Take notes about revisiting an activity later in the year.

- Take notes about modifying an activity for use with another foundation.

- Take notes about how you might leverage the activity in future years.

- Identify an activity that is particularly useful.

MODULE 1: NUMBER RELATIONSHIPS: COMPARISON AND ESTIMATION				
NO.	PAGE	TYPE	NAME	NOTES
1.1	23	T	Estimation Station	
1.2	24	T	Near and Far	
1.3	25	T	Where Does It Go?	
1.4	27	T	Open to Ticked	
1.5	29	T	Prompts for Relative Size of a Number	
1.6	30	R	"Dynamic Number Line"	
1.7	32	R	"The Stand"	
1.8	33	G	*Number Line Cross Off*	
1.9	35	G	*Five Targets*	
1.10	37	G	*Close To*	
1.11	38	G	*Ten Close Calls*	
1.12	40	C	The Sort	

MODULE 2: SUBITIZING AND DECOMPOSING				
NO.	PAGE	TYPE	NAME	NOTES
2.1	46	T	Dominoes Subitize and Sort	
2.2	47	T	Parts and Whole Models	
2.3	48	T	Break It, Move It, Prove It	
2.4	50	T	Express It	
2.5	52	T	Prompts for Subitizing and Decomposing	
2.6	53	R	"Quick Images in Color"	
2.7	54	R	"The Big Red Ten"	
2.8	55	G	*Finders Sitters*	
2.9	56	G	*Hot Number Potato*	
2.10	57	G	*Five Ways, Most Ways*	
2.11	58	C	Bag of Blocks: Predict and Break	

MODULE 3: DISTANCE TO 10, 100, AND 1,000				
NO.	PAGE	TYPE	NAME	NOTES
3.1	63	T	Ten Frames and Counters	
3.2	65	T	Cuisenaire Tens and Hundreds	
3.3	66	T	Are We There Yet?	
3.4	67	T	I Have, Who Has	
3.5	68	T	Prompts for Distance to 10, 100, and 1,000	
3.6	69	R	"Complex Number Strings (Combinations of 10, 100, and 1,000)"	
3.7	71	R	"How Many to Ten? With Quick Images"	
3.8	73	G	*Next Ten, Last Ten*	
3.9	75	G	*Combinations*	
3.10	77	G	*Parts and Stripes*	

MODULE 4: COUNTING AND SKIP COUNTING				
NO.	PAGE	TYPE	NAME	NOTES
4.1	81	T	30-Second Counts	
4.2	82	T	The Missing	
4.3	84	T	Dot Card Counting On	
4.4	85	T	Meter Stick Count Back	
4.5	86	T	Prompts for Counting and Skip Counting	
4.6	87	R	"The Count"	
4.7	88	R	"Stop the Count"	
4.8	89	G	*Crossing Over*	
4.9	91	G	*Number Catch*	
4.10	93	G	*300 Is Perfect*	
4.11	94	C	Shake and Spill	
4.12	95	C	Flip	

MODULE 5: PROPERTIES OF ADDITION AND ITS INVERSE RELATIONSHIP WITH SUBTRACTION				
NO.	PAGE	TYPE	NAME	NOTES
5.1	100	T	Twist It!	
5.2	101	T	Mingle and Match	
5.3	102	T	Which Switch?	
5.4	103	T	They're All Related!	
5.5	104	T	Prompts for Properties of Addition and Its Inverse Relationship With Subtraction	
5.6	105	R	"That's the Truth"	
5.7	107	G	*The Match*	
5.8	108	G	*Three Numbers for 10 Points*	
5.9	110	G	*Spoon Scramble*	
5.10	112	G	*Flip and Fill*	
5.11	114	C	Which Switch? (The Center)	
5.12	116	C	Triangle Cards	

MODULE 6: PROPERTIES OF MULTIPLICATION AND THE INVERSE RELATIONSHIP WITH DIVISION

NO.	PAGE	TYPE	NAME	NOTES
6.1	123	T	Commutative Frayer	
6.2	124	T	The Associative Prism	
6.3	126	T	You Have Your Work Cut Out for You	
6.4	127	T	Situations	
6.5	128	T	Which Is It?	
6.6	129	T	Prompts for Properties and Inverse Relationships	
6.7	130	R	"The Truth"	
6.8	132	G	*Multiplication Match*	
6.9	134	G	*Triangle Card Cover*	
6.10	135	C	Property Math Libs	

MODULE 7: MULTIPLYING BY TENS AND HUNDREDS				
NO.	PAGE	TYPE	NAME	NOTES
7.1	140	T	Multiples of Tens and Hundreds	
7.2	142	T	Groups of Disks	
7.3	144	T	Multiplying Tens With Cuisenaire Rods	
7.4	145	T	That's a Fact	
7.5	146	T	Prompts for Multiplying by Tens and Hundreds	
7.6	147	R	"Three Counts"	
7.7	149	R	"Complex Number Strings"	
7.8	150	G	*Six Charts*	
7.9	151	G	*Paper Clip Products*	
7.10	152	G	*For Keeps*	
7.11	154	G	*Multiples Bingo*	
7.12	155	C	Is the Same, Is the Same	

MODULE 8: MULTIPLES AND FACTORS				
NO.	PAGE	TYPE	NAME	NOTES
8.1	162	T	Building Multiples	
8.2	163	T	Breaks It, Makes It	
8.3	164	T	Multiples Venn Gallery Walk	
8.4	165	T	Building Tables of Special Multiples	
8.5	166	T	Prompts for Multiples and Factors	
8.6	167	R	"The Stand (Multiples Version)"	
8.7	168	R	"Knockout"	
8.8	169	G	*Multiple Cover Up*	
8.9	170	G	*Just Three Factors*	
8.10	172	G	*The Connects*	
8.11	173	G	*Five Card Multiples*	
8.12	175	G	*Dropping Primes*	

MODULE 9: DOUBLING AND HALVING				
NO.	PAGE	TYPE	NAME	NOTES
9.1	182	T	Cuisenaire Proofs	
9.2	183	T	Decompose and Double	
9.3	184	T	Breaking for Half	
9.4	186	T	Can You Halve an Odd Number?	
9.5	187	T	Prompts for Doubling and Halving	
9.6	188	R	"High/Low Doubles"	
9.7	189	R	"String of Halves"	
9.8	190	G	*For Keeps (Doubles or Halves)*	
9.9	191	G	*The Splits*	
9.10	192	G	*First to Seven*	
9.11	193	G	*50–150*	
9.12	194	C	Halve It Up	

MODULE 10: COMPUTATIONAL ESTIMATION				
NO.	PAGE	TYPE	NAME	NOTES
10.1	200	T	What Is Reasonable?	
10.2	202	T	Do I Need to Estimate?	
10.3	203	T	Just One	
10.4	204	T	Estimating With Worked Examples	
10.5	205	T	Prompts for Computational Estimation	
10.6	206	R	"Is It Reasonable?"	
10.7	207	R	"Over/Under"	
10.8	209	G	*Stay or Go*	
10.9	210	G	*Make It Close*	
10.10	212	G	*High/Low Estimates*	
10.11	214	G	*Chips*	
10.12	216	C	Make Five Estimates	
10.13	218	C	Sorting Estimates	
10.14	219	C	Just One, Which One	

References

Aguirre, J. M., Mayfield-Ingram, K., & Martin, D.B. (2013). *The impact of identity in K–8 mathematics: Rethinking equity-based practices.* National Council of Teachers of Mathematics.

Baroody, A. J., & Dowker, A. (Eds.). (2003). *The development of arithmetic concepts and skills: Constructing adaptive expertise.* Lawrence Erlbaum.

Bay-Williams, J. M., & Fletcher, G. (2017). A bottom-up hundreds chart? *Teaching Children Mathematics, 24*(3), 153–160

Bay-Williams, J. M., & SanGiovanni, J. J. (2021). *Figuring out fluency in mathematics teaching and learning, Grades K–8.* Corwin.

Berry, R. (2018, December). *Thinking about instructional routines in mathematics teaching and learning.* National Council of Teachers of Mathematics. https://www.nctm.org/News-and-Calendar/Messages-from-the-President/Archive/Robert-Q_-Berry-III/Thinking-about-Instructional-Routines-in-Mathematics-Teaching-and-Learning/

Bruner, J. S., & Kenney, H. J. (1965). Representation and mathematics learning. *Monographs of the Society for Research in Child Development, 30*(1), 50–59.

Chapin, S., O'Conner, C., & Anderson, N. (2013). *Classroom discussions: Using math talk to help students learn* (3rd ed.). Math Solutions.

Clements, D. H. (1999). Subitizing: What Is It? Why Teach It? *Teaching Children Mathematics, 5*(7), 400–405.

Flores, M. M., Burton, M., & Hinton, V. (2018). *Making mathematics accessible for elementary students who struggle: Using CRA/CSA for interventions.* Plural.

Franke, M. L., Kazemi, E., & Turrou, A. C. (2018). *Choral counting & counting collections: Transforming the PreK–5 math classroom.* Stenhouse Publishers.

Fuchs, L., Bucka, N., Clarke, B., Dougherty, B., Jacobson, J., Jordan, N., Karp, K., Weiss, C., & Woodward, J. (2021). U.S. Department of Education Institute of Science (IES) What Works Clearinghouse (WWC) practice guide on assisting students struggling with mathematics: Intervention in the elementary grades. Institute of Science.

Geary, D. C. (2011). Cognitive predictors of achievement growth in mathematics: A five-year longitudinal study. *Developmental Psychology, 47*(6), 1539–1552.

Griffin, C. C., Jossi, M. H., & van Garderen, D. (2014). Effective mathematics instruction in inclusive schools. In J. McLeskey, N. L. Waldron, F. Spooner, & B. Algozzine (Eds.), *Handbook of effective inclusive schools.* Routledge.

Hiebert, J., & Grouws, D. A. (2007). The effects of classroom mathematics teaching on students' learning. In F. K. Lester (Ed.), *Second handbook of research on mathematics teaching and learning* (pp. 371–404). Information Age Publishing.

Huinker, D. (2013). Dimensions of fraction operation sense. In *Defining mathematics education: Presidential yearbook selections 1926–2012, seventy-fifth yearbook of the National Council of Teachers of Mathematics (NCTM)*, pp. 373–380. National Council of Teachers of Mathematics.

National Council of Teachers of Mathematics (NCTM). (2014). *Principles to actions: Ensuring mathematical success for all.* NCTM.

National Council of Teachers of Mathematics (NCTM). (2023). *Procedural fluency in mathematics: Reasoning and decision-making, not rote application of procedures position.* https://www.nctm.org/uploadedFiles/Standards_and_Positions/Position_Statements/PROCEDURAL_FLUENCY.pdf

National Research Council. (2001). *Adding it up: Helping children learn mathematics.* J. Kilpatrick, J. Swafford, and B. Findell (Eds.). Mathematics Learning Study Committee, Center for Education, Division of Behavioral and Social Sciences and education. National Academy Press.

SanGiovanni, J. J. (2020). *Daily routines to jump-start math class: Elementary school*. Corwin.

SanGiovanni, J. J. (2023). *Daily routines to jump-start problem solving: K–8*. Corwin.

SanGiovanni, J., Katt, S., & Dykema, K. (2020). *Productive math struggle: A six-point action plan for fostering perseverance*. Corwin.

SanGiovanni, J., Katt, S., Knighten, L., & Rivera, G. (2022). *Answers to your biggest questions about teaching elementary math*. Corwin.

Star, J. R. (2005). Reconceptualizing conceptual knowledge. *Journal for Research in Mathematics Education, 36*(5), 404–411.

Van de Walle, J. A., Karp, K. S., & Bay-Williams, J. M. (2023). *Elementary and middle school mathematics: Teaching developmentally* (11th ed.). Pearson Education.

Index

**JENNIFER M. BAY-WILLIAMS,
JOHN J. SANGIOVANNI, ROSALBA SERRANO,
SHERRI MARTINIE, JENNIFER SUH,
C. DAVID WALTERS, SUSIE KATT**

Because fluency is so much more than
basic facts and algorithms.
Grades K–8

**MARIA DEL ROSARIO ZAVALA,
JULIA MARIA AGUIRRE**

Discover innovative equity-
based culturally responsive
mathematics instruction that
unlocks the mathematical
heart of each student.
Grades K–8

**IVANNIA SOTO,
THEODORE RUIZ SAGUN,
MICHAEL BEIERSDORF**

Focus on the literacy
opportunities that multilingual
students can achieve when
language scaffolds are
taught alongside rigorous
math and science content.
Grades K–8

**JOHN J. SANGIOVANNI, SUSIE KATT,
LATRENDA D. KNIGHTEN, GEORGINA RIVERA,
FREDERICK L. DILLON, AYANNA D. PERRY,
ANDREA CHENG, JENNIFER OUTZS, KAREN MESMER,
ENYA GRANDOS, KEVIN GANT, LAURA SHAFER**

Actionable answers to your most pressing
questions about teaching elementary math,
secondary math, and secondary science.

Elementary, Secondary

**CHRISTA JACKSON, KRISTIN L. COOK,
SARAH B. BUSH,
MARGARET MOHR-SCHROEDER,
CATHRINE MAIORCA, THOMAS ROBERTS**

Help educators create integrated STEM
learning experiences that are inclusive for all
students and allow them to experience STEM
as scientists, innovators, mathematicians,
creators, engineers, and technology experts!

Grades PreK–5 and Grades 6–12

CM23841706

A Sage Company

CORWIN HAS ONE MISSION: to enhance education through intentional professional learning.

We build long-term relationships with our authors, educators, clients, and associations who partner with us to develop and continuously improve the best evidence-based practices that establish and support lifelong learning.